W9-CMZ-991

Energy and Power

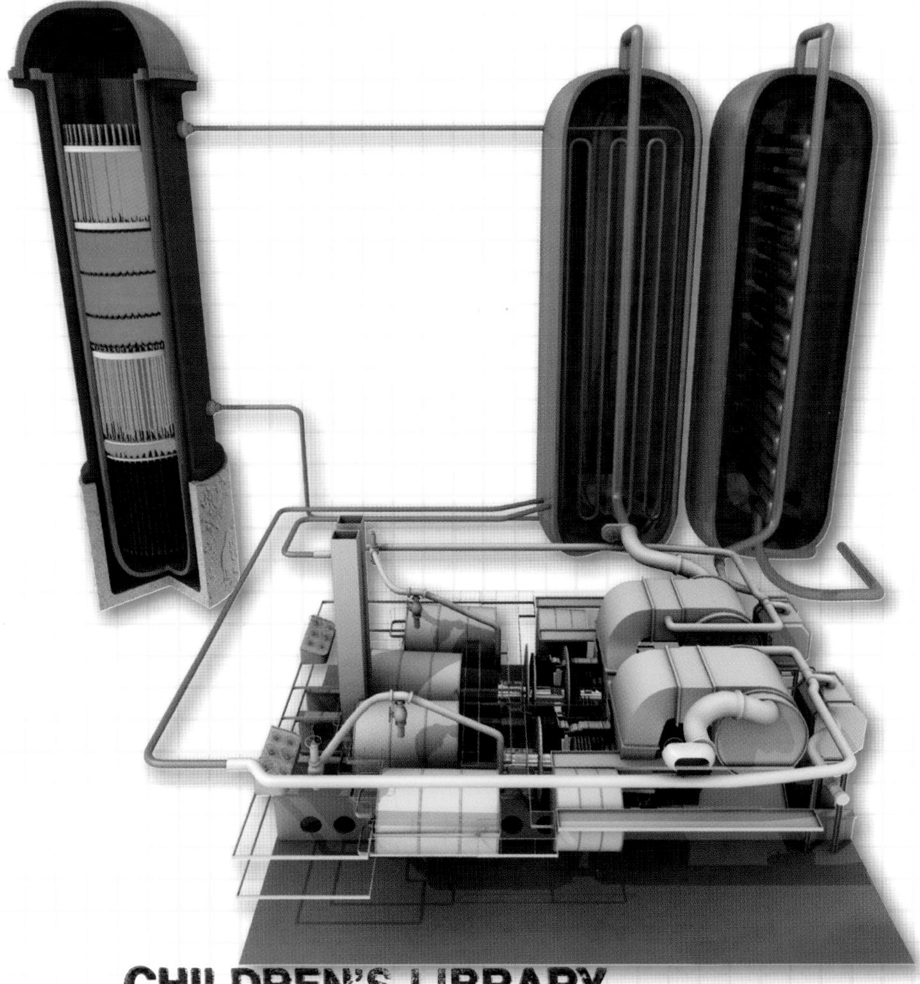

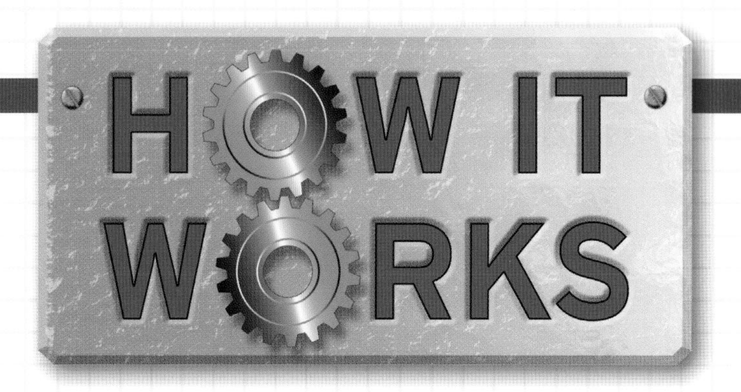

Energy and Power

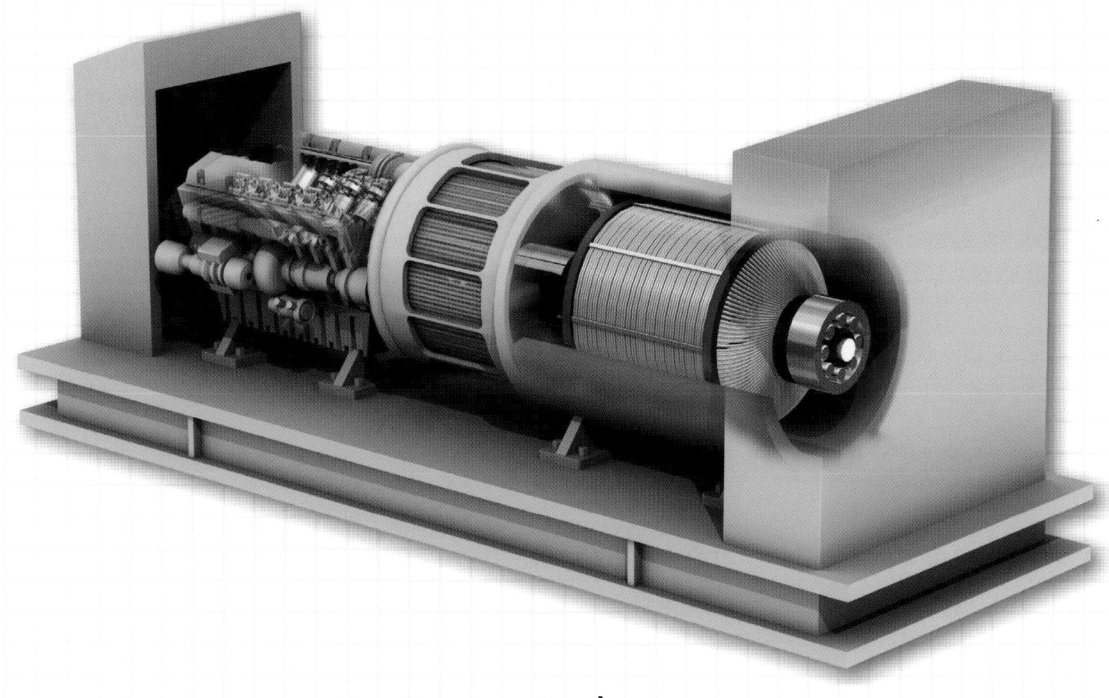

By Steve Parker
Illustrated by Alex Pang

MASON CREST PUBLISHERS INC.
370 Reed Road, Broomall, Pennsylvania 19008
(866)MCP-BOOK (toll free), www.masoncrest.com

First Printing
9 8 7 6 5 4 3 2 1

Library of Congress Cataloging-in-Publication Data
Parker, Steve, 1952–
Energy and power / by Steve Parker ; illustrated by Alex Pang.
p. cm. — (How it works)
Includes bibliographical references and index.
ISBN 978-1-4222-1794-8
Series ISBN (10 titles): 978-1-4222-1790-0
1. Power resources—Juvenile literature. 2. Electric power plants—Juvenile literature. I. Pang, Alex, ill. II. Title.
TJ163.23.P375 2011
621.31—dc22
2010033616
Printed in the U.S.A.

First published in 2009 by Miles Kelly Publishing Ltd
Harding's Barn, Bardfield End Green, Thaxted, Essex, CM6 3PX, UK

Editorial Director: *Belinda Gallagher*
Art Director: *Jo Brewer*
Design Concept: *Simon Lee*
Volume Design: *Rocket Design*
Cover Designer: *Simon Lee*
Indexer: *Gill Lee*
Production Manager: *Elizabeth Collins*
Reprographics: *Stephan Davis, Ian Paulyn*
Consultants: *John and Sue Becklake*
Assets Manager: *Bethan Ellish*

Every effort has been made to acknowledge the source and copyright holder of each picture. Miles Kelly Publishing apologizes for any unintentional errors or omissions.

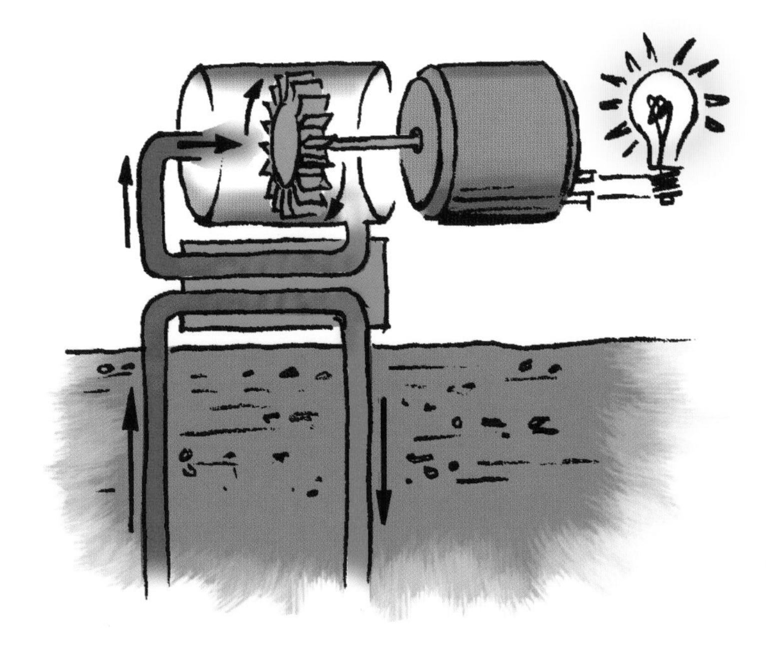

ACKNOWLEDGEMENTS

All panel artworks by Rocket Design
The publishers would like to thank the following sources for the use of their photographs:
Alamy: 25 John Novis; 34 greenwales
Corbis: 26 Joseph Sohm/Visions of America; 28 Harald A. Jahn; 36 Kim Kulish
Fotolia: 13 Lottchen
Getty Images: 19 Boris Horvat; 21 Harald Sund
Photolibrary: 7 (c) Manfred Bail
Rex Features: 11 Sipa Press; 30 Jon Santa Cruz
Science Photo Library: 17 US Department of Energy; 33 Andrew Lambert Photography
All other photographs are from Miles Kelly Archives

WWW.FACTSFORPROJECTS.COM

Each top right-hand page directs you to the Internet to help you find out more. You can log on to **www.factsforprojects.com** to find free pictures, additional information, videos, fun activities and further web links. These are for your own personal use and should not be copied or distributed for any commercial or profit-related purpose.

If you do decide to use the Internet with your book, here's a list of what you'll need:
- A PC with Microsoft® Windows® XP or later versions, or a Macintosh with OS X or later, and 512Mb RAM
- A browser such as Microsoft® Internet Explorer 7, Firefox 2.X or Safari 3.X
- Connection to the Internet via a modem (preferably 56Kbps) or a faster Broadband connection
- An account with an Internet Service Provider (ISP)
- A sound card for listening to sound files

Links won't work?
www.factsforprojects.com is regularly checked to make sure the links provide you with lots of information. Sometimes you may receive a message saying that a site is unavailable. If this happens, just try again later.

Stay safe!
When using the Internet, make sure you follow these guidelines:
- Ask a parent's or a guardian's permission before you log on.
- Never give out your personal details, such as your name, address or email.
- If a site asks you to log in or register by typing your name or email address, speak to your parent or guardian first.
- If you do receive an email from someone you don't know, tell an adult and do not reply to the message.
- Never arrange to meet anyone you have talked to on the Internet.

The publisher is not responsible for the accuracy or suitability of the information on any website other than its own. We recommend that children are supervised while on the Internet and that they do not use Internet chat rooms.

CONTENTS

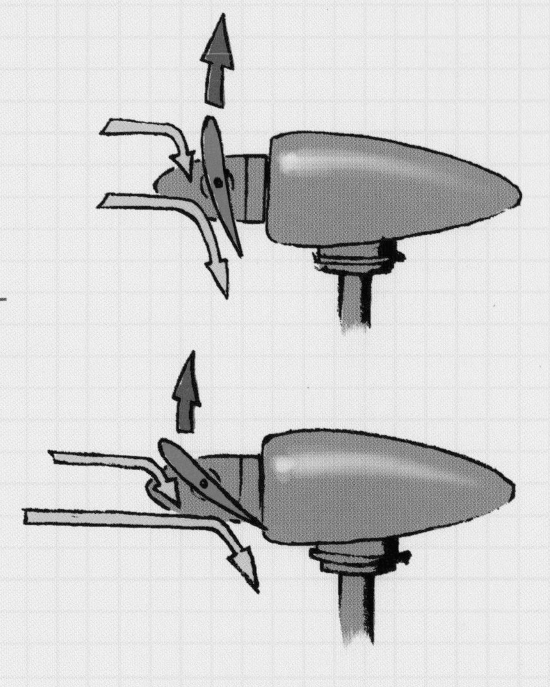

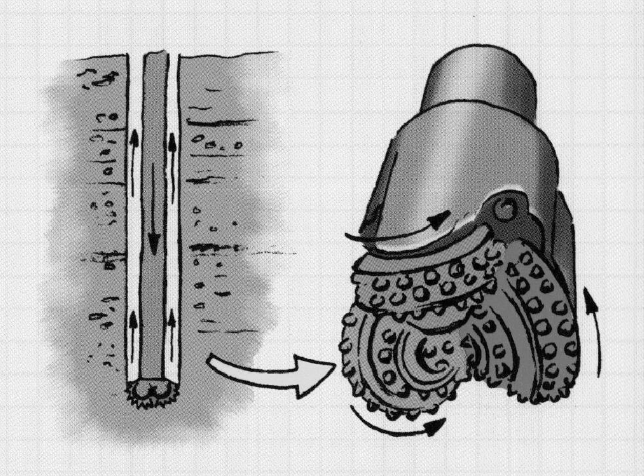

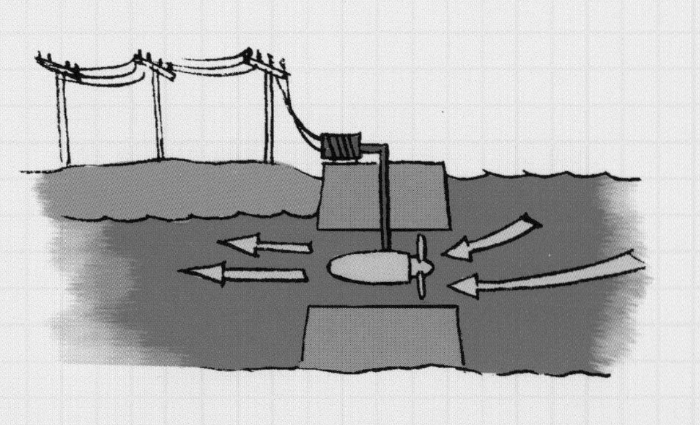

INTRODUCTION

Energy comes in many forms—movement, heat, light, sound, chemical substances in fuels, electrical and magnetic forces, and radioactivity. We use energy when we flick on a light switch, cook a meal or travel in a plane. Energy is never created or destroyed—it changes from one form to another. For example, a car's fuel energy changes into heat, sound and movement. In a television, electrical energy becomes tiny dots of colored light.

Energy from fast-flowing water is used to generate electricity at a hydroelectric power station.

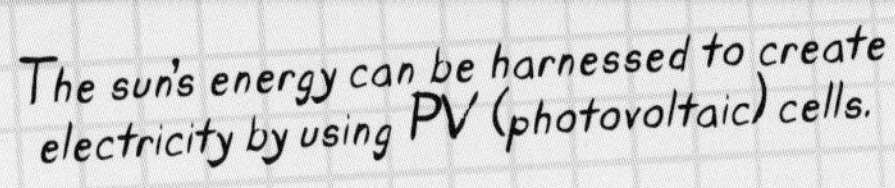
The sun's energy can be harnessed to create electricity by using PV (photovoltaic) cells.

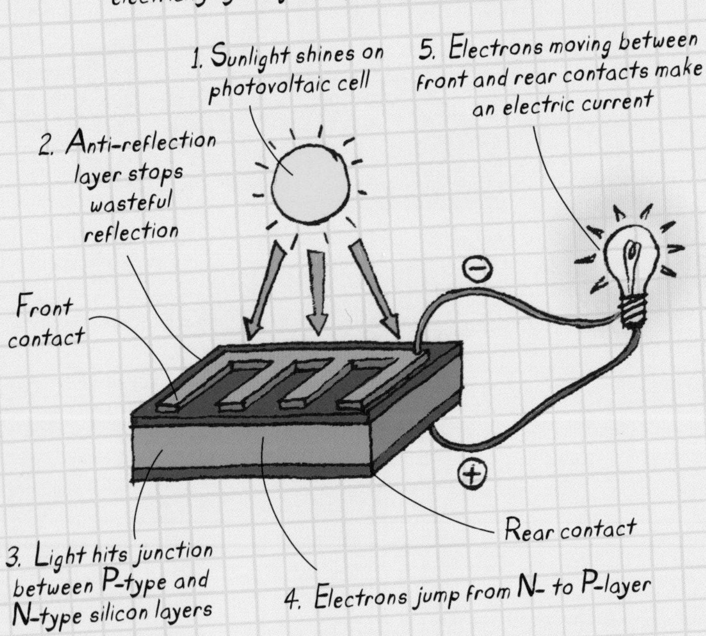
1. Sunlight shines on photovoltaic cell
2. Anti-reflection layer stops wasteful reflection
3. Light hits junction between P-type and N-type silicon layers
4. Electrons jump from N- to P-layer
5. Electrons moving between front and rear contacts make an electric current
Front contact
Rear contact

SUN POWER

Our world runs on energy changes. It started when ancient people burned wood, turning its chemical fuel energy into heat for cooking and warmth. Next came movement or kinetic energy—air blowing on a windmill's sails, and flowing water pushing a watermill's paddles, to grind grain or lift water from wells. These forms of energy can all be traced back to the sun. Its light is captured by trees to grow wood. Its heat warms air to cause winds, and turns water into rising water vapor, which turns back into water droplets in clouds, falls as rain and flows downhill into rivers.

FOSSIL ENERGY

Our major energy sources today also come from the sun—or rather, they did. Coal, oil and gas are fossil fuels. Coal is the preserved remains of plants from millions of years ago. Petroleum oil and natural gas come from preserved tiny sea plants and animals. These fuels provide five-sixths of all energy worldwide, especially for industry, vehicles, heating and electricity-generating power stations.

Giant oil platforms satisfy our fossil fuel needs—for now.

All of the pages in this book are Internet linked. Visit www.factsforprojects.com to find out more.

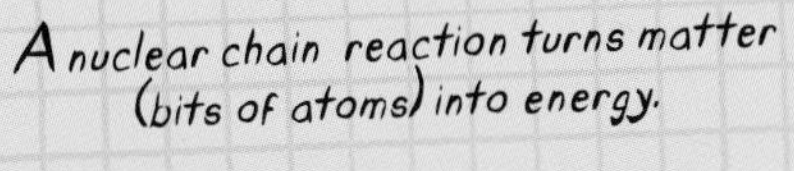
A nuclear chain reaction turns matter (bits of atoms) into energy.

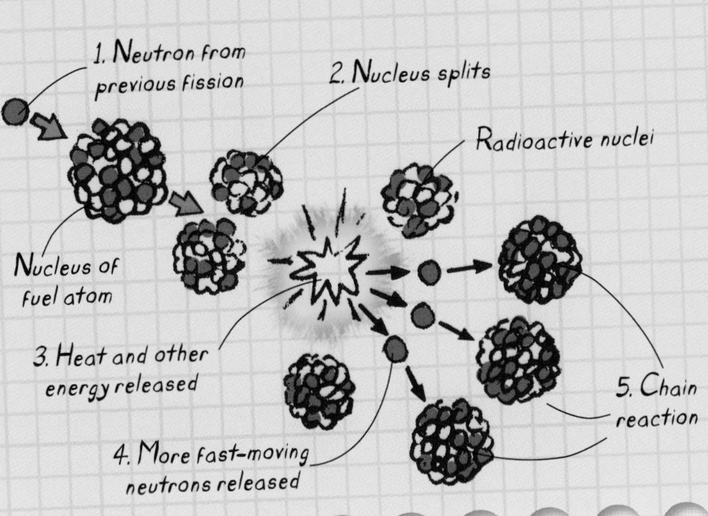

ENERGY CRISIS

If we continue using fossil fuels at today's rate, they will probably run out in less than 200 years. Also, burning them makes greenhouse gases that speed up global warming and worsen climate change. The same goes for burning biofuels from plants such as wood, straw and oil, and biogas from rotting decay. There is little effect on global warming from a nuclear power station, once it's up and running. However it has other hazards, such as the risk of accidents and radioactive products that no one can make truly safe.

The solar stove goes back to basics, focussing light from the sun to boil water and cook food.

LOW-E-FUTURE

There are solutions to the energy crisis. We can reduce energy needs with more efficient transport, heating systems, lighting and insulation. Cleaner, greener, sustainable and renewable energy sources—hydro (flowing water), solar (sun), wind, tides, waves and geothermal (heat from deep underground) can also be used.

Fifty years from now people may look back at our energy waste and wonder: "Didn't they care?"

The eco-friendly Subaru G4e has almost zero emissions and runs on electricity—G4e stands for "Great for Earth."

COAL MINE

About 250 years ago, coal was the main fuel driving the steam engines of the Industrial Revolution. Today it is the main fuel burned worldwide to generate electricity. Over the years, new sources of coal were discovered, mined in quantity and gradually used up. So mines and coal-burning power stations shifted around the world, and still do.

Eureka!

Small-scale coal mining dates back more than 2,000 years to Roman times. Early large-scale mining began in the north-east of England, around Durham and Newcastle, from the 1700s.

What next?

There is enough coal in the world to last for about 150 years. But coal-burning power stations are a major cause of global warming.

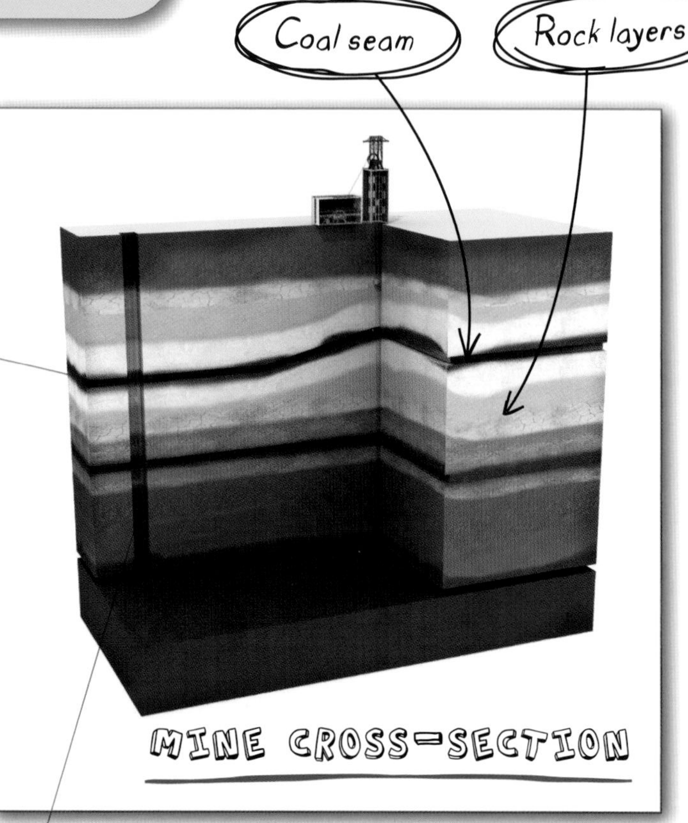

Tunnels The excavated tunnels follow the layers or seams of coal, angling up or down as necessary.

Direction of mining

Longwall shearer moves to and fro across coal face

Self-advancing roof supports

Conveyer belt removes coal

Some areas of roof are allowed to collapse (known as the gob area)

Ventilation Shafts to the surface and powerful fans remove dangerous explosive gases and heat from the tunnels and replace them with fresh air.

The world's greatest coal producer is China, closely followed by Russia, then India, Australia and South Africa.

How do COAL FACE CUTTERS work?

About half of underground coal is cut by longwall mining. A machine called a longwall shearer moves to and fro across a wide face or wall of coal. Its rotating drum and sharp metal teeth scrape and loosen the coal into lumps. These fall onto a conveyor belt or a chain of large bucket- or pan-like containers. The belt or buckets carry the raw cut coal to the vertical conveyor or coal elevator. As the shearer eats its way forwards, supports or props hold up the ceiling of rock behind it for as long as needed. Then the ceiling is allowed to fall in or collapse.

Winding room The large reel or drum, driven by electric motors, winds in the cables to raise the elevator cages. It lets out the cables in a safe, controlled manner to lower the cages.

For a virtual tour of a coal mine visit www.factsforprojects.com and click on the web link.

Headframe Also called the winding tower, this is usually built of steel girders or concrete beams to hold up the pulley gear.

Over the centuries a belief grew up that carrying a small lump of coal brings good luck—you will always have something to keep you warm.

Lifting cables
Strong steel cables, hundreds or even thousands of metres long, lift the elevator cages for the miners and equipment.

Headgear pulleys
The cables run over a set of large pulleys (winding wheels) to raise and lower the cages from the pithead to the tunnels far below. Bigger pulleys mean easier winding and less bending strain on the cable.

Elevator shaft

Blasting coal from a surface mine

Surface strip mining

Coal is usually removed from open-cast or open-cut (surface) mines by the strip method. A long strip is drilled and the holes filled with explosives. A loud siren warns of the blast, which blows up the coal into small lumps. These are removed by excavators, draglines, dump-trucks or conveyors. Some coal conveyor belts are more than 9 miles (15 km) long.

About two-thirds of the world's coal is mined underground. The rest comes from open-cast mining of exposed seams at the surface, or those easily uncovered by removing a thin overlying rock layer known as overburden.

The oldest, hardest and most energy-packed type of coal is anthracite, formed up to 400 million years ago. Lignite is much softer, burns cooler, and may be less than two million years old.

OIL AND GAS PLATFORM

Some of the biggest structures in the world are oil and gas platforms. They are used anywhere there's petroleum oil or gas below—these two energy-packed fossil fuels often occur together. The principle of drilling is simple, as the toothed bit grinds its way deep into rocks. But it's a rough, tough process with constant dangers including fires, explosions and high-pressure blowouts.

Eureka!

The most famous early oil strike was by Edwin Drake near Titusville, Pennsylvania, USA in 1859. However other oil boreholes were drilled before this, such as the Bibi-Eibat well of 1848 near the city of Baku, Azerbaijan, by Russian engineer FN Semyenov.

At sea, exploratory oil rigs or platforms make the first boreholes to see if the area has any oil or gas. After this early work, a production platform may be brought to the site to extract the oil or gas over many years.

One of the world's tallest structures is the Petronius oil platform in the Gulf of Mexico. It stands 2,001 ft (610 m) above the ocean floor.

Accommodation block Like a small town, the platform has many amenities for the workers including bedrooms, kitchens, canteen, gymnasium, cinema and other leisure areas. There are also workshops for maintenance and repair jobs, and laboratories for scientific testing of rock, oil and gas samples.

Crane Large onboard cranes unload food, water, scientific equipment and other stores from supply ships that moor alongside the platform. The rig must have enough supplies to last out several weeks of stormy weather when ships cannot reach it.

How does the DRILL BIT work?

Few machines have such a hard-wearing time as the oil drill bit. Its three toothed wheels interlock or mesh so that they rasp and bite through solid rock as the drill string spins. A specially made fluid "mud" is pumped down the hollow inside of the drill string. It cools and lubricates the bit wheels, and flows back up around the drill string, carrying pieces or cuttings of rock. At the surface the expensive 'mud' is filtered and pumped down again.

Drill string is extended at the top as the bit drills
Waste rock is carried up around the string
Casing
Bit and borehole are wider than drill string
Toothed wheels made of toughened metal rotate to scrape off and crush rock

Decks The busiest areas are the wide, flat main decks, which have the rotary table (turntable) and its electric motors.

Legs Some rigs are fixed, with legs that are flooded with water to stand on the ocean floor. Others have hollow legs and partially float, kept in position by cables or stays anchored to the seabed.

Take a tour of a drilling rig, view it from different angles and click on the parts to see what they do by visiting www.factsforprojects.com and clicking on the web link.

Some test drillings have gone down more than 29, 527 ft (9,000 m) without finding oil or gas.

Derrick This tall tower has pulleys and cables that raise each new length of drill pipe so that it can be added to the drill string, section by section.

Kelly The kelly is a piece of pipe with four or six flat sides that is screwed into the top of the drill string, and turned around by the rotary table.

Rotary table Powered by an electric motor, this clamps onto the kelly and makes it turn around to twist the drill string and the bit far below.

Helipad Helicopters bring new crews and urgent supplies every few days. They take away many items, from an ill or injured worker to the latest samples of oil, gas and seabed rock.

Upper deck

Lower deck

The energy in oil and gas is measured in a unit called the BOE (Barrel of Oil Equivalent). One BOE is about 1.7 MWh (megawatt hours). That's enough energy to supply a typical house with electricity for four months.

The Hibernia platform on the Jeanne d'Arc Basin, in the North Atlantic Ocean near Newfoundland, weighs more than 40,786 tons (37,000 metric tons). It has drilled more than 50 boreholes in the seabed around it.

Longer and deeper

The drill string is a long chain of drill pipes screwed together end to end. The whole drill string turns around, driven by the rotary table on the platform. To lengthen the string, a new section of pipe is hoisted up inside the tower or derrick, and screwed into the end of the uppermost pipe below. The whole string is then lowered to begin spinning again.

Clamps hold the drill pipes so that they can be attached

POWER STATION

There are about ten main kinds of power (electricity-generating) stations. They all convert or transform one source of energy into another—which is electricity. In fossil fuel power stations the energy source is coal, oil or gas. It is burned to make heat energy, which is converted using high-pressure steam and turbines into movement (kinetic) or mechanical energy, which spins the generators to make electricity (see page 14).

Eureka!

The first large-scale power generator for paying customers was the Pearl Street Station in New York City, USA. Built by the Edison Electric Light Company, it used coal as fuel and started in 1882 with about 85 customers.

Cooling tower Excess heat is removed in these giant towers. Newer power stations often lack cooling towers. Instead the heat is used to warm nearby buildings and to provide hot water.

The world's biggest coal-fuelled power stations include the Kendal Power Station near Johannesburg, South Africa, and the Yuhuan Power Station in Zhejiang Province, China. They produce more than 4,000 MW (megawatts), enough electricity for half a million people.

Boilers Pulverized coal (coal ground into tiny pieces) or another fossil fuel burns here continuously. As one boiler shuts down for maintenance or repair, the others keep going.

Fuel Coal is fed continuously from hoppers (temporary storage containers) and ash is removed.

Each set of turbine blades is specially shaped and angled to get the most energy from the steam at that position

Steam leaves at much lower speed and pressure

Spinning shaft is connected to generator

High-pressure steam blasts into turbine intake

How do TURBINES work?

A turbine is a device that converts the flow of a fluid, such as superheated steam or water, into a spinning motion. It has long, slim blades fixed onto a central shaft in a radial pattern, like the spokes of a wheel. Most turbines have several sets of blades. Their sizes and angles are carefully designed to extract the most energy from the steam, which loses speed and pressure as it moves past each set of blades.

Find out how electricity is generated and take a virtual tour of a power station by visiting www.factsforprojects.com and clicking on the web link.

Thomas Edison's Pearl Street Station lost money for the first few years, partly because there was no existing electricity network. The company had to install all the wiring to the customers—the first distribution grid.

Distribution grid The electric current is taken away along thick metal cables called power lines. These join into the larger electricity network known as the distribution grid.

Pylons Tall towers hold the high-voltage power lines well above the ground.

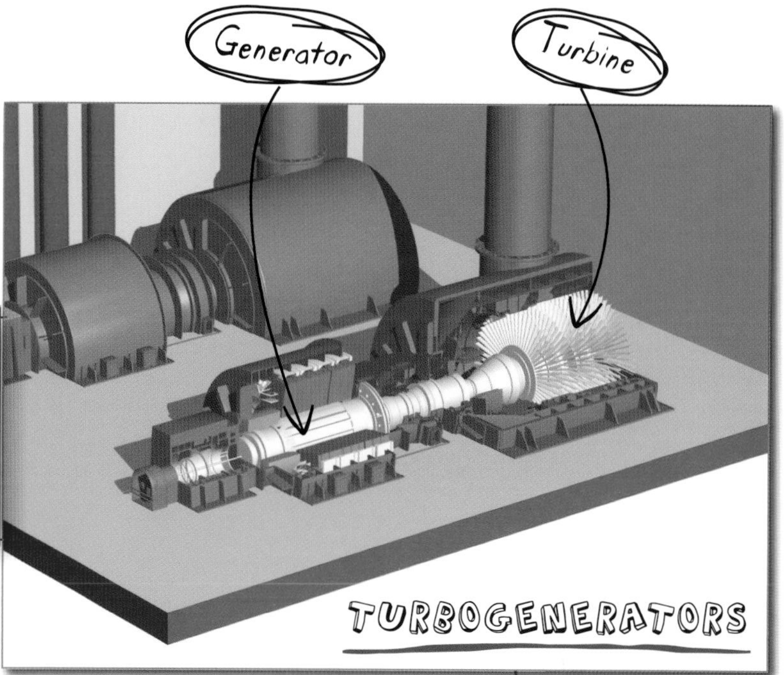

Generator

Turbine

TURBOGENERATORS

Steam return

Steam feed

Transformers

Turbogenerators In the turbine hall, multi-bladed turbines convert the continuous blast of superheated steam into a spinning motion to generate electricity (see pages 14–15).

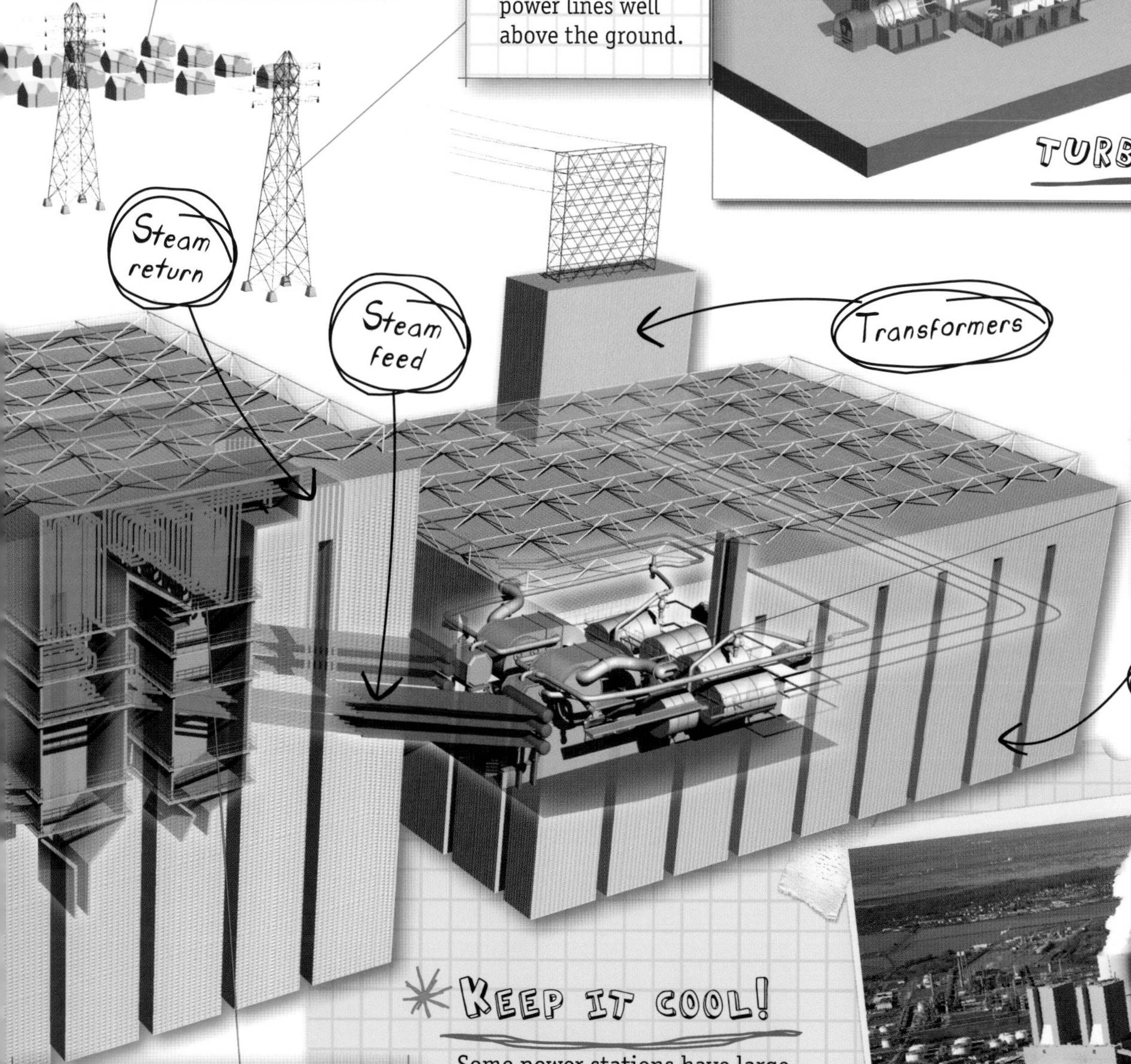

Turbine hall

Steam circuit Boilers superheat water into high-pressure steam that is led along sets of pipes to the turbines. The steam is then cooled back into water and recycled to the boilers.

KEEP IT COOL!

Some power stations have large cooling towers where spare heat is lost. The hot water gives off invisible water vapor. As this rises into cooler air it condenses back into clouds of water droplets or "steam."

Cooling towers make ground-level "clouds" of water droplets

ELECTRICITY GENERATOR

The heart of a power station is the generator, which changes mechanical or kinetic energy (the energy of movement) into electrical energy. The mechanical energy is usually in the form of a rotary or spinning motion. This is harnessed directly from moving wind or water, or from the superheated steam produced by burning fossil fuels such as coal, oil and gas, or biofuels like wood—even animal droppings!

Eureka!

Generators depend on an effect called electromagnetic induction. This was discovered in about 1831 by English scientist and experimenter Michael Faraday, and at around the same time, but separately, by US scientist and engineer Joseph Henry.

What next?

The working temperature of a generator affects how much electricity it produces. Supercooled materials of the future could make the process more efficient, giving more electricity for less fuel.

Current take-off contacts

Casing

Some power station transformers deal with half a million volts

Transform it!

An electric current's pushing strength is measured in volts. Most power station generators make currents of a few thousand volts. But electricity travels best over long distances in power lines at hundreds of thousands of volts, because it loses less energy as heat and in other ways. Power station step-up transformers increase the voltage from the generators to 400,000 volts or more for the power lines. At the other end, step-down transformers reduce it for daily use, for example, to 240 volts in Britain.

Small generators driven by petrol engines can provide electricity at the normal household voltage almost anywhere in the world. There are even underwater models!

Stator The outer sets of wire coils (which stay still) make up the stator. Electricity fed into them from the grid makes a magnetic field around them by the electromagnetic effect.

Watch an interactive animation showing a generator in action by visiting www.factsforprojects.com and clicking on the web link.

Main shaft The rotor spins on a long shaft between high-precision bearings. It must keep turning at a steady speed to make sure the voltage output stays constant.

Most generators make AC (alternating current). The electricity flows one way and then the other many times each second (for example, 50 times per second in Britain).

Rotor The inner sets of wire coils that spin around are known as the rotor. As they turn within the magnetic field made by the stator, electricity is induced to flow through them.

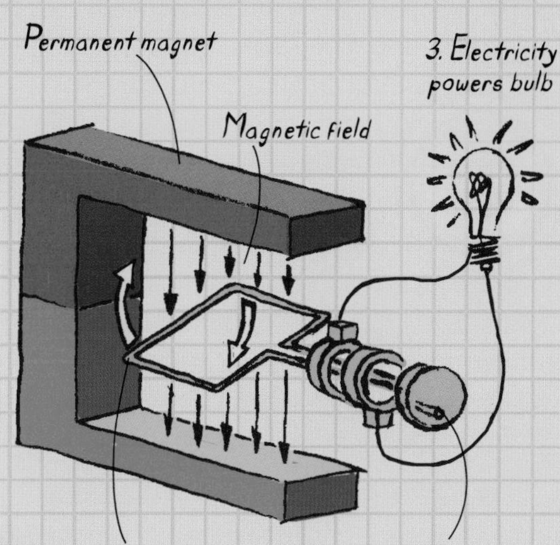
Permanent magnet
3. Electricity powers bulb
Magnetic field
2. Electricity is induced in wire coil rotor spinning in magnetic field
1. Mechanical energy turns wire coil rotor

Stator magnetic field
Main shaft ball bearings

How do GENERATORS work?

Electricity and magnetism make up one of the Universe's four basic forces—electromagnetism. If a wire moves near a magnet, electricity flows along it—this is electromagnetic induction. In a simple generator, a wire coil (the rotor) spins within the magnetic field of a permanent magnet made of iron. In a power station generator, the permanent magnet is replaced by another set of wire coils, the stator. As electricity passes through them they make a magnetic field by the electromagnetic effect. This is when a flowing electric current produces a magnetic field around itself. It's the "opposite" of electromagnetic induction.

Windings Each of the rotor and stator coils has thousands of turns of wire, to produce as much electricity as possible. The windings are carefully designed so that their individual magnetic fields add together and do not interfere or cancel out, which would waste energy.

Because of the spinning motion of the generator, it produces AC rather than DC (direct current) which flows steadily one way only.

Hungarian scientist and priest Anyos Jedlik started experimenting with early generators in the 1820s. He made a small working version in the 1850s. Jedlik also invented carbonated or fizzy drinks.

NUCLEAR POWER STATION

Like most power stations, the nuclear generating plant has steam turbines, electricity generators and transformers. Its special feature is the way it produces the heat energy to boil the water for the turbines. It has a nuclear reactor where heat is made by splitting apart nuclei—the central parts of atoms. Atoms are the smallest pieces or particles of ordinary substance or matter.

Eureka!

The first nuclear chain reaction was made at the University of Chicago, USA in 1942. The team, led by Enrico Fermi, built the reactor, called a "pile," in a converted racket court under the university's sport grandstand.

What next?

In the USA, radioactive nuclear waste was due to be buried deep underground at a site called Yucca Mountain, Nevada. However in 2009, after 22 years of planning, the idea was dropped.

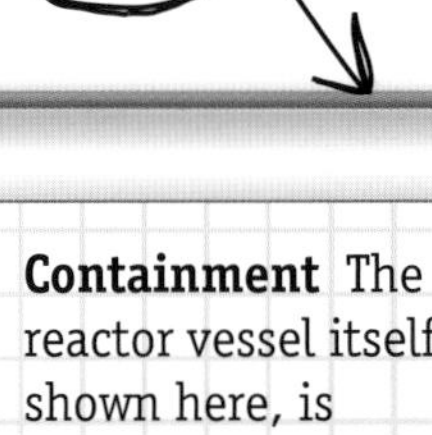

Fuel rods The nuclear fuel is usually made in the shape of long rods, which can be lowered into the main part of the reactor, surrounded by the moderator.

Containment The reactor vessel itself, shown here, is housed in an outer casing designed to keep in, or contain, any radioactive gases and other substances accidentally released.

What is NUCLEAR FISSION?

A typical nuclear power station uses nuclear fission—the nuclei are broken apart (as opposed to nuclear fusion, see page 18). The fuel is usually a metal such as uranium or plutonium, whose huge atoms come apart easily. External energy starts the reaction, breaking up the nucleus of a fuel atom. This gives off fast-moving particles called neutrons, and these do the same to nearby nuclei, producing heat, radioactivity and other forms of energy as they go. The key is to control the chain reaction so it does not get out of hand and explode.

1. Neutron from previous fission smashes into nucleus
2. Nucleus splits
3. Heat and other energy is released
4. More fast-moving neutrons released
5. Neutrons smash into more fuel atom nuclei

Nucleus of fuel atom

Radioactive nuclei

A nuclear chain reaction

Control rods These rods take in, or absorb, excess neutron particles to make sure the chain reaction does not get out of hand. They can be raised or lowered to adjust the progress of the reaction.

Moderators The moderator substance slows down the neutron particles so that they produce continual heat rather than a sudden explosion. Moderators include water, graphite (a form of carbon) and, more rarely, the metal-like substance beryllium.

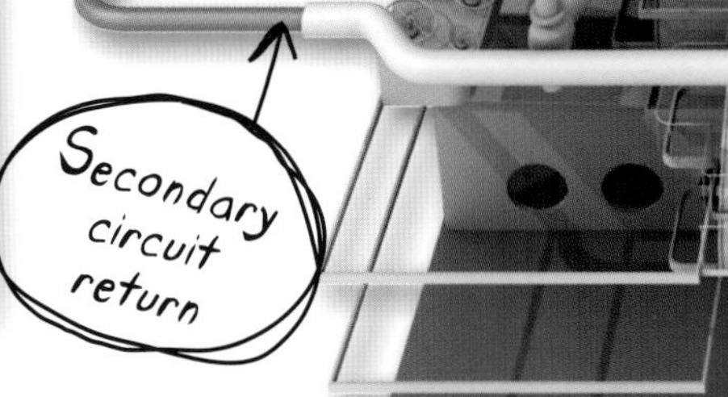

See how a chain reaction happens by visiting www.factsforprojects.com and clicking on the web link.

Radioactive waste

Nuclear reactions produce radioactive waste of used or spent fuel, fluids, pipes and other equipment. These give off harmful rays and particles known as radiation. Some of this waste will be dangerous for thousands of years. For now, the waste is stored in deep ponds or concrete containers. Some of the fuel is moved to reprocessing plants to be treated and used again.

Primary circuit Fluid circulates between the reactor and the heat exchangers inside this sealed circuit or loop.

These giant radioactive waste tanks will be encased in concrete and earth

Secondary circuit Heat from the reactor is brought here by the primary circuit. It boils water into superheated steam that flows through the secondary circuit pipes to the turbines and back.

The biggest nuclear accident was at Chernobyl, Ukraine, in 1986. Reactor Number 4 exploded and released clouds of radioactive gases and dust.

The world's biggest nuclear power station is Japan's Kashiwazaki-Kariwa Nuclear Power Plant. Its seven reactors generate 8,000-plus MW (megawatts).

Heat exchangers Heat energy passes from the primary circuit to the secondary circuit in these numerous loops of pipes, keeping any radioactivity in the primary circuit.

Secondary circuit feed

At Chernobyl, about 50 people died at the scene. More than 4,000 died over the following years from the after-effects, as gases and dust spread across Europe and exposed millions of people to low-level radiation. In some areas even cows' milk was too radioactive to drink.

Steam turbogenerators The high-pressure steam from the secondary circuit spins turbines that turn generators, as in most other thermal (heat-based) power stations.

FUSION POWER

Most nuclear power stations generate electricity from heat made by breaking apart big nuclei, known as nuclear fission. Experimental reactors are testing fusion power, where small nuclei are brought together and joined or fused into bigger ones, giving off heat in the process.

Eureka!

Fusion power is well known in nature. The Sun and other stars make their immense amounts of light and heat using it. The idea for a fusion reactor to generate electricity was suggested by George Thompson and Moses Blackman as long ago as the 1940s.

What next?

So far, fusion experiments have used up more energy than they produced. ITER, an international reactor being built in Cadarache, France, should produce more energy than it uses when it starts up around 2015–2020.

The Sun's natural fusion power happens in its centre or core, where the temperature is 27 million degrees Fahrenheit (15 million degrees Celsius). Here, hydrogen nuclei fuse together in the gigantic heat and pressure to make helium.

Experimental fusion reactors include JET, the Joint European Torus (a torus is a doughnut-shaped ring) at Culham, England, and KSTAR in Korea. KSTAR's magnets weigh 330 tons (300 metric tons).

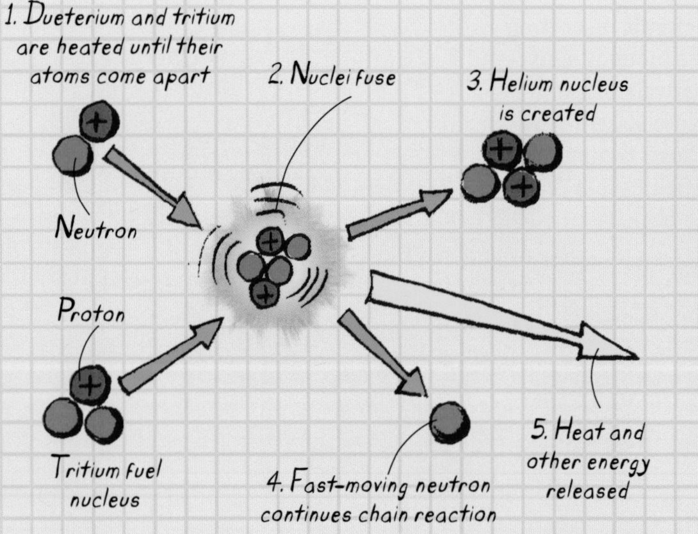
1. Dueterium and tritium are heated until their atoms come apart
2. Nuclei fuse
3. Helium nucleus is created
Neutron
Proton
Tritium fuel nucleus
4. Fast-moving neutron continues chain reaction
5. Heat and other energy released

Toroidal magnets The incredibly hot plasma inside the reactor is kept trapped or confined by magnetic fields from several sets of electromagnets. The outer toroidal magnets make a doughnut-shaped field.

Tokomak The main part of the reactor is a doughnut-shaped vessel called a tokomak. Sets of huge wire coils around it work as electromagnets.

Fuel inlet

How does FUSION work?

Fusion fuel is deuterium or tritium or a combination of both. These substances are naturally occurring heavier forms of hydrogen, the lightest substance with the smallest, simplest atoms. The fuel is heated to millions of degrees, into a form or state of matter called plasma. Its atoms come apart and two of their nuclei join or fuse to make a bigger nucleus of the substance helium. As this happens, a fast-moving neutron is given off and fuses two more nuclei, and so on, in a chain reaction. The plasma is so hot it would melt any substance it touches. So it must be held in place in the reactor by intense magnetic fields.

Read more facts about fusion power and find out facts about ITER by visiting www.factsforprojects.com and clicking on the web link.

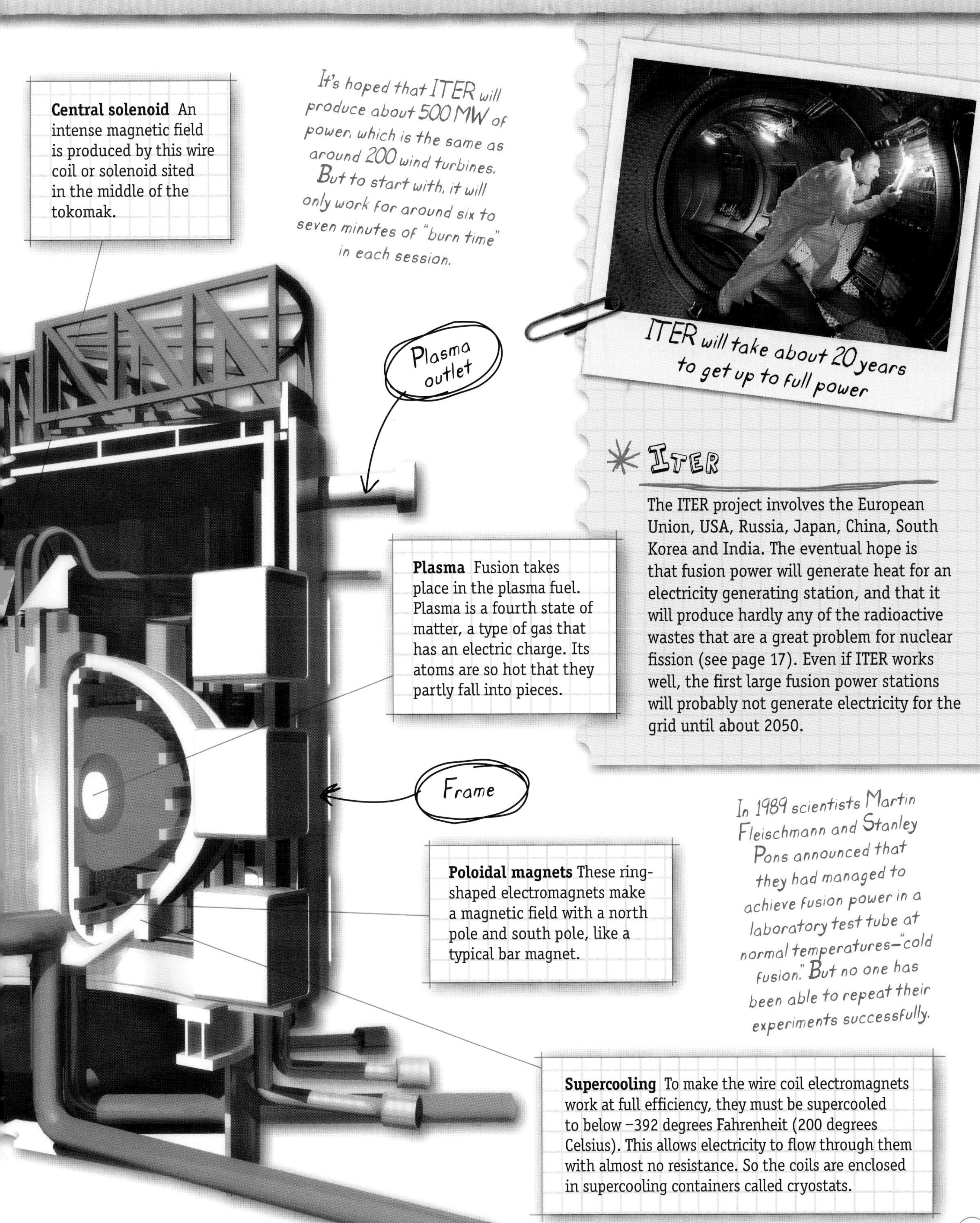

Central solenoid An intense magnetic field is produced by this wire coil or solenoid sited in the middle of the tokomak.
Plasma outlet
Plasma Fusion takes place in the plasma fuel. Plasma is a fourth state of matter, a type of gas that has an electric charge. Its atoms are so hot that they partly fall into pieces.
Frame
Poloidal magnets These ring-shaped electromagnets make a magnetic field with a north pole and south pole, like a typical bar magnet.
Supercooling To make the wire coil electromagnets work at full efficiency, they must be supercooled to below −392 degrees Fahrenheit (200 degrees Celsius). This allows electricity to flow through them with almost no resistance. So the coils are enclosed in supercooling containers called cryostats.

It's hoped that ITER will produce about 500 MW of power, which is the same as around 200 wind turbines. But to start with, it will only work for around six to seven minutes of "burn time" in each session.

ITER will take about 20 years to get up to full power

ITER

The ITER project involves the European Union, USA, Russia, Japan, China, South Korea and India. The eventual hope is that fusion power will generate heat for an electricity generating station, and that it will produce hardly any of the radioactive wastes that are a great problem for nuclear fission (see page 17). Even if ITER works well, the first large fusion power stations will probably not generate electricity for the grid until about 2050.

In 1989 scientists Martin Fleischmann and Stanley Pons announced that they had managed to achieve fusion power in a laboratory test tube at normal temperatures—"cold fusion." But no one has been able to repeat their experiments successfully.

HYDROELECTRIC POWER STATION

Unlike fossil fuel and nuclear power stations, the hydroelectric generating plant does not have a direct heat source. Its energy comes from water flowing along a river. The water runs downhill under the force of gravity, having fallen as rain far upstream. That rain was once water in clouds, and it was originally evaporated from the sea and lakes by the heat of the Sun. So a hydroelectric power station is, in effect, Sun-powered.

Eureka!

An early use of flowing water for energy was the waterwheel. This was common in ancient Greece, more than 2,000 years ago, to turn millstones and grind grain into flour, and to raise river water up into ditches for crops.

What next?

Hydroelectricity is increasing around the world since it does not involve burning fuels that contribute to global warming. Norway produces 99 percent of its electricity in this way.

The Itaipu Complex of dams on the Parana River between Brazil and Paraguay is 4.5 miles (7.2 km) long and produces 14,000 MW.

Sluice gates and spillways If the river floods and too much water builds up behind the dam, sluice gates or flood gates are opened to let some of it flow away downstream.

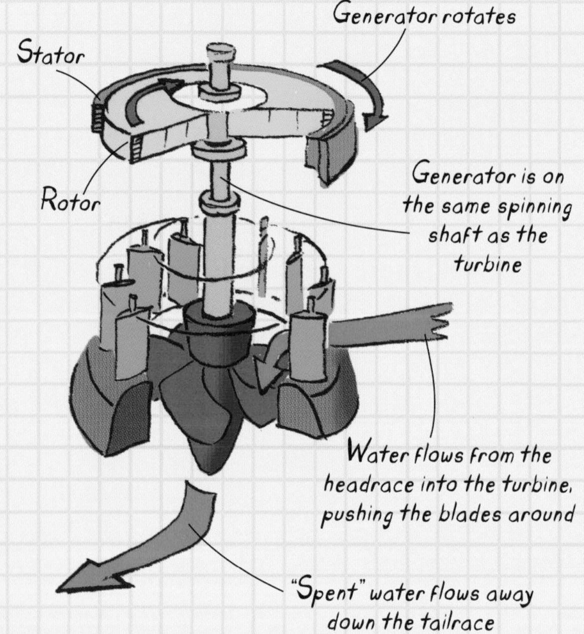
Generator rotates
Stator
Rotor
Generator is on the same spinning shaft as the turbine
Water flows from the headrace into the turbine, pushing the blades around
"Spent" water flows away down the tailrace

Dam The dam is in effect a wall across the river, holding back the water on the upstream side. It is thicker at the base to cope with increased water pressure at depth.

Tailrace This pipe or duct carries the water away from the turbine, through the base of the dam to the downstream section of the river.

Downstream flow

How does HYDROELECTRICITY work?

The hydroelectric turbine has angled blades that are spun by the force of water pushing past them. The water is made to build up behind the dam to give it increased pressure or pushing force, known as the height or "head" of water. This body of water is also a store or reserve for times when it rains less and the river level falls, giving it the name reservoir.

The first hydroelectric power station was built in 1882 on the Fox River in Wisconsin, USA—just after the Pearl Street Station, New York.

Watch an animation to show the effects of dams on rivers by visiting www.factsforprojects.com and clicking on the web link.

Reservoir Water trapped behind the dam builds up into an artificial lake called a reservoir. This can be very useful for supplying water to farmers' fields—and for swimming, sailing, fishing and other leisure pursuits.

Control vanes The angle of these vanes or slats is adjusted to keep the flow of water steady, depending on its pressure, which in turn depends on the surface level of the reservoir.

Shaft

Generator The whole turbine set-up must be in a waterproof casing to keep water away from the generator just above.

Thrust bearing This huge ball-bearing takes up the stresses and pressure of the water so that they do not affect the electricity generator.

Headrace (penstock) This pipe or duct takes the water from the inlets to the turbines. Grilles or meshes over the inlets keep out large objects.

TURBINE

Blades The turbine blades are shaped to take as much movement energy as possible from the water as it flows past.

Inlet

Turbines

DAM BYPASS

Hydroelectric plants do not produce greenhouse gases or radioactive wastes. But they do alter a river's flow, flood the upstream area with a reservoir and take water away from downstream regions, all of which greatly alters local wildlife. The dam also blocks the way for animals that journey or migrate along the river. Some dams have a series of pools at the side. Fish such as salmon can jump from one pool to the next, a small distance each time, to get past the dam.

The "fish ladder" acts as a dam bypass

The world's largest power stations are hydroelectric. The Three Gorges Complex on the Chang-Jiang (Yangtze) River in China generates 22,000 MW.

GEOTHERMAL ENERGY

At its center or core, the Earth is exceedingly hot—probably more than 9,032 degrees Fahrenheit (5,000 degrees Celsius). It gradually cools to an average of just 59 degrees Fahrenheit (15 degrees Celsius) at the surface. Even though we cannot drill very deep, we can still get down to the hot rocks and bring their heat energy to the surface, to warm our buildings and water, and to generate electricity.

Eureka!

The first geothermal energy plants to provide heating for homes and buildings began in the 1890s, in places as far apart as Paris, France and Boise, Idaho, USA. The earliest attempts to generate electricity from naturally heated steam were made at Larderello, Italy, in the 1900s. Japan followed in 1919 and California, USA in 1921.

What next?

Experiments are taking place to see if it is possible to "tap" runny lava, the red-hot molten (liquid) rock from volcanoes, and use its heat to produce electricity.

On average, rock temperatures increase by 96.1 °F (35.6 °C) for every 328 ft (100 m) of depth. So after about one half of a mile (1 km) down, it can be 68 degrees hotter than at the surface. No wonder deep mine tunnels get so warm!

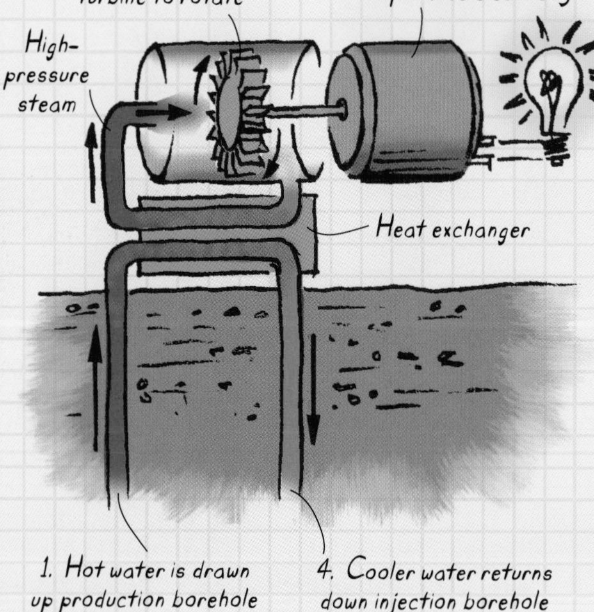
2. Steam pressure causes turbine to rotate
3. Turbine spins generator to produce electricity
High-pressure steam
Heat exchanger
1. Hot water is drawn up production borehole
4. Cooler water returns down injection borehole

How does GEOTHERMAL ENERGY work?

Heat from inside the Earth is carried up to the surface by a liquid substance or medium, commonly water, in the primary circuit. As the water warms, it rises naturally, gets to the surface, passes its heat energy to water in the secondary circuit, cools and is pumped back down again. In the secondary circuit, water or a similar "flash vapor" liquid boils and drives the turbogenerators.

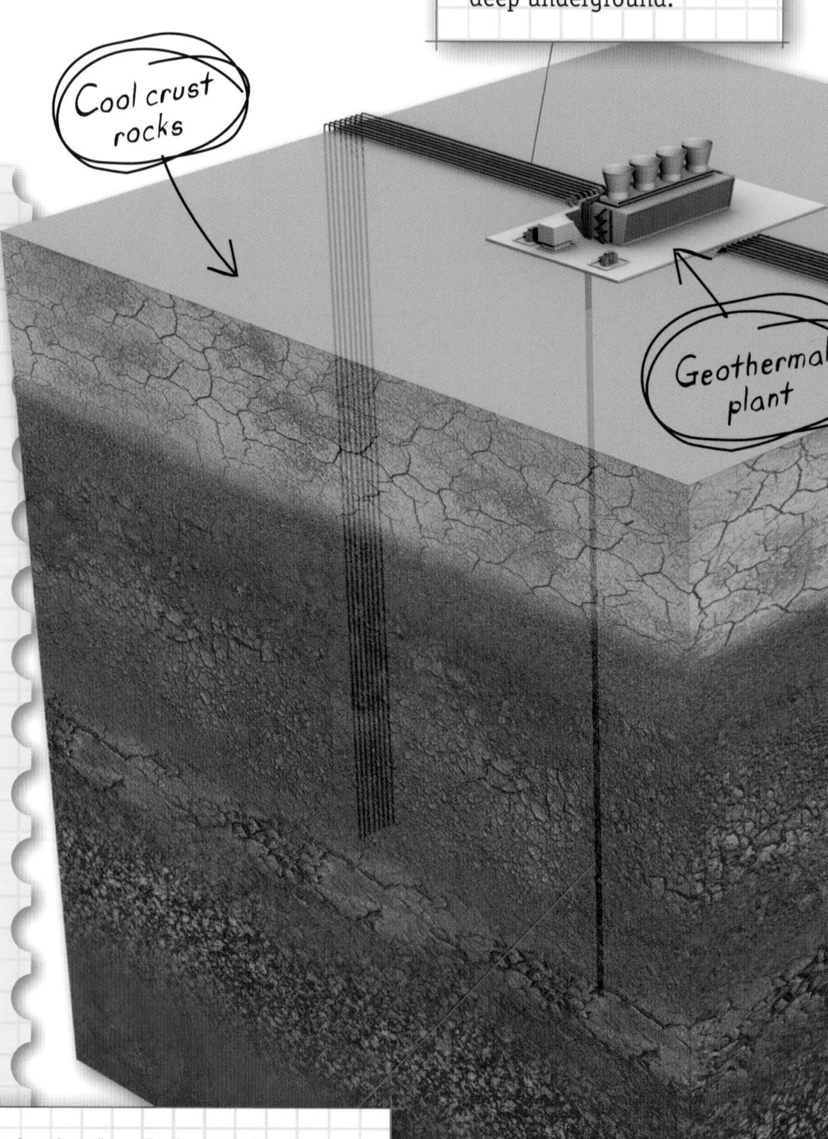
Cool crust rocks
Geothermal plant

Production (extraction) boreholes Pipes carry up very hot, high-pressure water and steam from deep underground.

Injection boreholes Water is pumped down at great pressure to replenish the water and steam taken up the production boreholes, otherwise the deep rock layers would gradually dry out.

To discover lots more information about geothermal energy visit www.factsforprojects.com and click on the web link.

In parts of Iceland, New Zealand and Japan, hot rocks are just below the surface. These regions are best for development of geothermal energy.

Hot water and steam blast from a natural geyser

THE GEYSER BLOWS!

We can see the Earth's natural heat energy at work when a geyser erupts. Deep below the geyser is a network of narrow holes and cracks in the hot rocks. Water trickles into them from the rocks around and becomes heated. Its pressure and temperature build up until they are high enough to make the water spurt up to the surface. The pressure is released, then more water starts to trickle in, and the whole process repeats, usually at regular intervals.

Rock strata (layers) The boreholes or wells pass through many layers of rock until they get to heat-bearing strata.

Cooling exhausts Excess heat is lost or vented to the air through cooling towers.

GEOTHERMAL PLANT

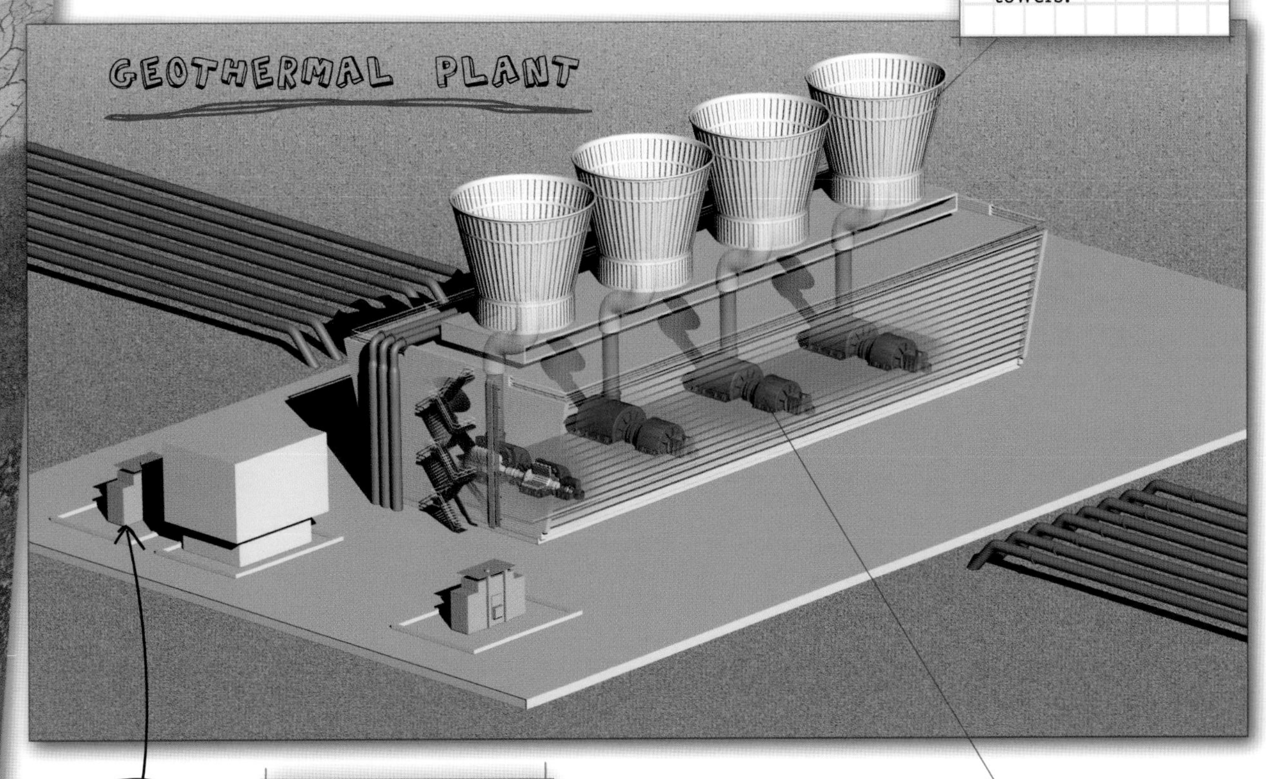

Magma Far below the surface the rock is so hot and pressurized that it is partly melted and known as magma. This is what erupts from volcanoes as red-hot runny lava.

Much of the heat inside the Earth is left over from when the planet formed out of a gas cloud. The cloud squeezed together under its own gravity, about 4.6 billion years ago.

Generators The turbines spin generators as in other types of power stations. The biggest geothermal plants produce up to 1,000 MW.

TIDAL BARRAGE

As the Earth spins around every 24 hours, the Moon pulls with its force of gravity on the seas and oceans. The Moon causes the water to 'bulge' towards it, and as the Earth slowly turns beneath, the bulge travels around the planet. We see the bulge as the raised sea level of a high tide, followed by a low tide as that part of the Earth carries on spinning away from the Moon. The energy in this vast and predictable water movement can be harnessed as tidal power for electricity.

What next?

Waves are caused by winds, due to the Sun's heating effect on the atmosphere (see page 26). But capturing wave power is difficult since big storms can easily destroy even the toughest equipment.

Inspection walkway

Gantry crane

A tidal barrage and its machinery must be cleaned regularly to avoid build-up of seaweeds and encrusting animals such as mussels, barnacles and limpets.

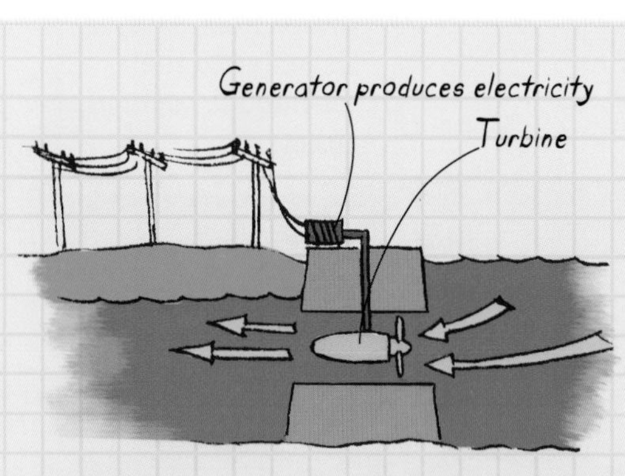
Generator produces electricity
Turbine
Tide comes in

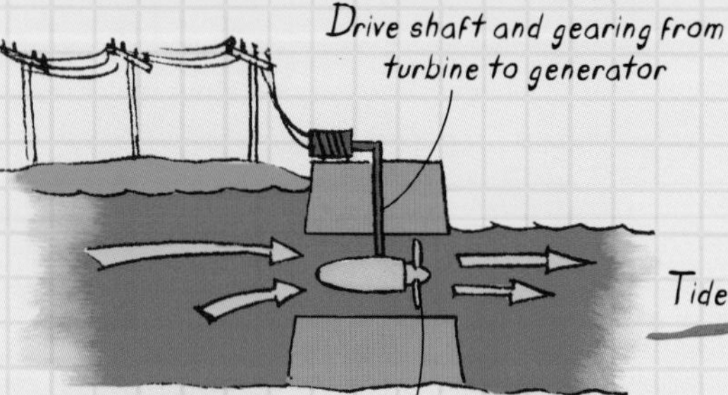
Drive shaft and gearing from turbine to generator
Tide goes out
Blades are spun by water and work in either direction

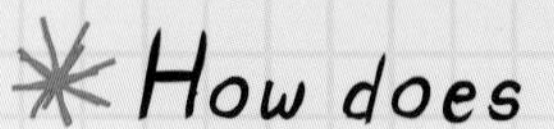

How does TIDAL POWER work?

Some tidal turbines are bi-directional. This means the blades spin one way as the tide comes in and water forces its way past the blades, from the open sea into the bay or estuary (river mouth). Then as the tide goes out and the sea level falls, the flow reverses and the blades spin the other way. Electrical circuits make sure the electricity current stays the same.

Generators The electricity generators (see page 14) convert the spinning motion of the turbines into electric current.

Turbines The turbines are in ducts that channel the water with greatest force over their blades.

Shrouded turbines, with the blades inside a collar-like shroud, are being tested on Australia's Gold Coast and Canada's west coast. The shrouds increase the speed of the water through the turbine.

Read facts about tidal barrages and lagoons and discover other green sources of energy by visiting www.factsforprojects.com and clicking on the web link.

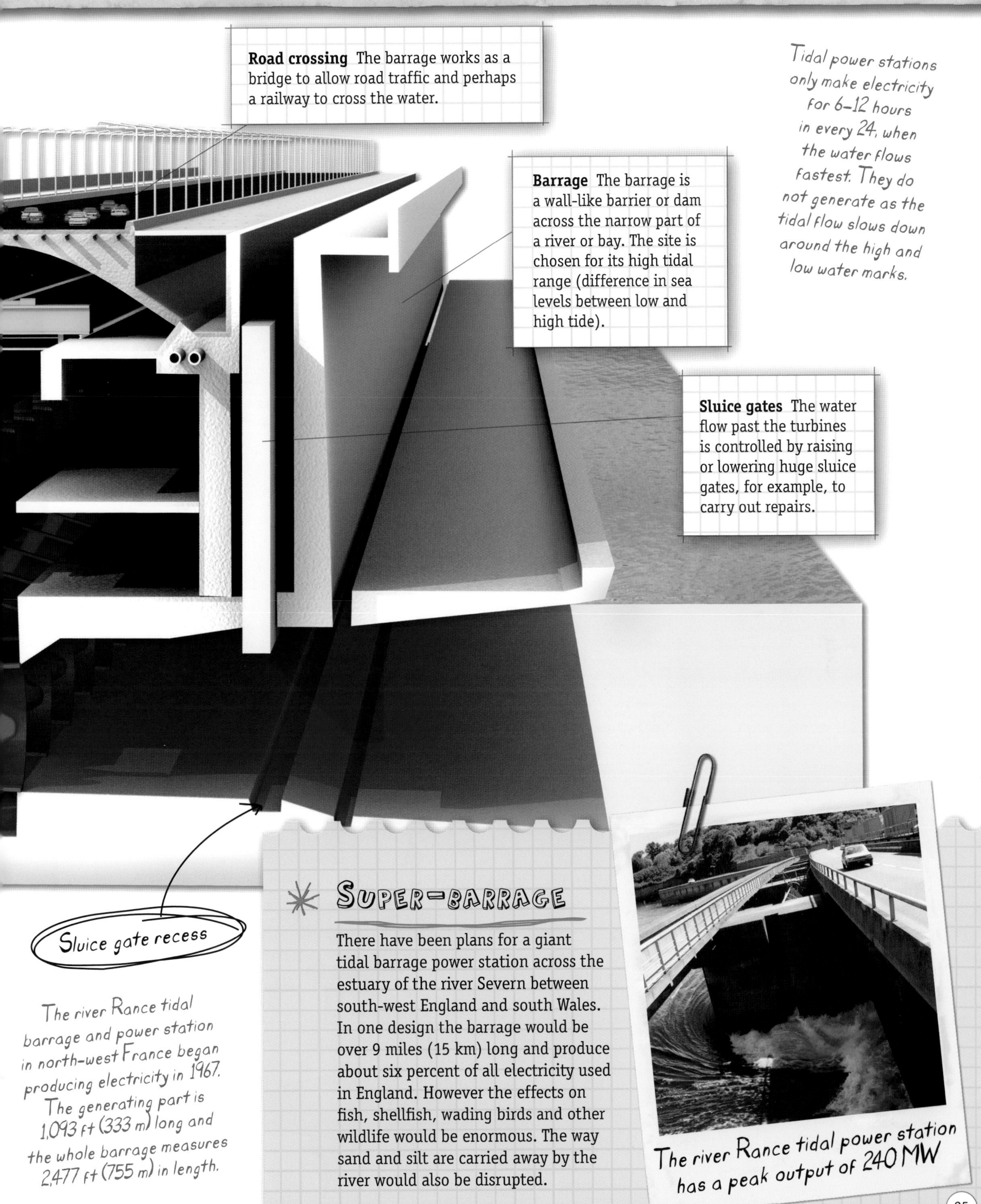

Road crossing The barrage works as a bridge to allow road traffic and perhaps a railway to cross the water.
Barrage The barrage is a wall-like barrier or dam across the narrow part of a river or bay. The site is chosen for its high tidal range (difference in sea levels between low and high tide).
Sluice gates The water flow past the turbines is controlled by raising or lowering huge sluice gates, for example, to carry out repairs.
Sluice gate recess

Tidal power stations only make electricity for 6–12 hours in every 24, when the water flows fastest. They do not generate as the tidal flow slows down around the high and low water marks.

The river Rance tidal barrage and power station in north-west France began producing electricity in 1967. The generating part is 1,093 ft (333 m) long and the whole barrage measures 2,477 ft (755 m) in length.

Super-barrage

There have been plans for a giant tidal barrage power station across the estuary of the river Severn between south-west England and south Wales. In one design the barrage would be over 9 miles (15 km) long and produce about six percent of all electricity used in England. However the effects on fish, shellfish, wading birds and other wildlife would be enormous. The way sand and silt are carried away by the river would also be disrupted.

The river Rance tidal power station has a peak output of 240 MW

WIND TURBINES

Like hydroelectricity and solar power, wind energy comes from the Sun. Its heat warms different parts of the land, sea and air by varying amounts. As hotter air rises, cooler air moves along to take its place—and this moving air is the wind blowing.

Eureka!

Wind power is not new. More than 1,200 years ago windmills in the Middle East turned millstones to grind grains into flour, and lifted water out of rivers up into the irrigation ditches for crops.

A huge wind turbine in Emden, Germany has rotor blades 413 ft (126 m) across and produces 7 MW of electricity.

Spinner

Blade

Pylon The tall tower holds the rotor blades high above ground level, so that they can rotate safely where the wind is stronger. Steps inside the tower allow engineers to get up to the machinery in the pod.

Darrieus turbines are a vertical axis design

New designs

Engineers test new wind turbine designs to see if they work better in very light or strong winds. In the Darrieus version the main shaft or axis of the turbine is upright. This vertical axis design reduces strain on the bearings compared to the design where the rotor shaft is parallel with the ground, known as horizontal axis.

How does variable pitch work?

To produce a regular flow and voltage of electricity, and avoid wind damage, the generator should turn at a constant speed. So the rotor blades change their angle or pitch. In high wind they swivel more edge-on to the moving air. The pushing force on them is less, which prevents their spin speed rising. In weak wind the blades twist more flat-on to the wind, for more turning force, to prevent their spin speed reducing.

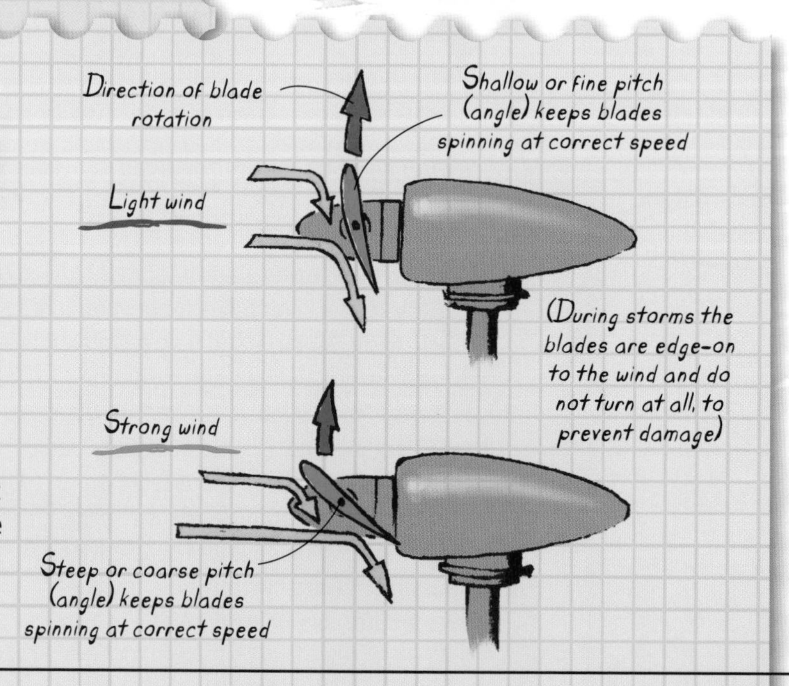
Direction of blade rotation
Shallow or fine pitch (angle) keeps blades spinning at correct speed
Light wind
(During storms the blades are edge-on to the wind and do not turn at all, to prevent damage)
Strong wind
Steep or coarse pitch (angle) keeps blades spinning at correct speed

Discover everything there is to [illegible] by visiting www.factsforprojects.[illegible]

What next?

A gyromill is a wind turbine on a long cable. Its whirling rotor blades generate electricity that comes down the cable, and the blades also work like helicopter rotors to keep the gyromill up.

Wind turbines are also called aerogenerators or wind power units. Lots of them in a windy place are known as a wind farm.

Once built and installed, wind turbines do not produce greenhouse gases. However they can be noisy and a hazard to birds, and spoil the view in beautiful scenery.

Gearbox The gearing system changes the slow rotation of the rotor blades to a faster spinning motion suitable for the generator.

Pitch control The angle or pitch of the rotor blades changes according to the speed of the wind.

Hub Strong bearings cope with enormous stresses as they allow the rotor shaft to spin in the hub.

Generator The mechanical spinning movement from the rotor is changed into electrical energy (see page 14) which travels along cables, down the pylon and away underground.

Some turbines have to be stopped at certain times because the sunlight flashing rapidly off their shiny blades can affect people nearby, causing headaches and even fits or convulsions.

In most wind turbines the blades are designed to spin about once every three to four seconds.

Blade The rotor blades are usually made of combined fibreglass-reinforced plastic (GRP) or a similar light, strong composite material.

Yaw control The pod swings around or yaws on a swivel bearing, to keep the rotor blades pointing into the wind.

Blade shape The blade surface moves through the air much more slowly near the hub compared to the tip. So the blade shape alters along its length, with the best angle and curve to get the most energy from the wind at that place.

Wind-generated electricity is far from constant, so other forms of generation will always be needed—or giant batteries to store the electricity.

BIOMASS ENERGY

Material from living things is known as biomass. Many types of biomass can be used as fuel for energy, either directly or indirectly. Wood, straw and other plant products are burned directly for warmth, cooking and to generate electricity. Rotting plants, animal bodies, droppings and other biomass decay to produce biogases such as methane, which can also be burned to provide heat or electricity.

Some power stations are designed to burn chicken droppings—plentiful where chickens are raised in huge sheds for eggs and meat.

Eureka!

Simple biogas digesters or fermenters were used in ancient times in the Middle East, India and China. They were built from stone slabs and covered by wooden poles, branches, twigs and soil. The basic design is still the same in modern versions, although the building materials have changed.

What next?

More and more waste disposal landfill sites have pipes to collect the methane given off by rotting leftover food, vegetable peelings, garden cuttings and similar materials. This is partly for energy and partly for safety, to prevent the methane exploding!

Incinerators

Incineration, or burning thoroughly at very high temperatures, is one way to get rid of rubbish that might otherwise go to landfill sites. It also produces great amounts of heat, typically burning at more than 1,472 degrees Fahrenheit (800 degrees Celsius), to warm buildings or generate electricity. This technology is known as WtE, waste-to-energy. However incinerating mixed rubbish usually gives off poisonous fumes and leaves behind toxic ash waste. Cleaning the fumes before they go out into the air, and disposing of the ashes, can cause considerable problems.

Incinerators are heat sources for generating electricity

Biogas Digester

Inlet Leftover food, plant matter from cooking and gardening, animal wastes and effluent—all together known as the feedstock—are added through the inlet. There is usually a "gas-trap" to prevent smells escaping.
Fixed dome The strong dome-shaped cover is rigid enough to prevent anything that moves over it above ground from falling into the tank below.
Biogas Most biomass rots anaerobically (without oxygen) to give off a mix of methane (about two-thirds), also carbon dioxide and small traces of nitrogen, ammonia, sulphur dioxide and hydrogen sulphide. This last gas gives a "rotten eggs" smell.

For lots of facts and diagrams about biomass energy visit www.factsforprojects.com and click on the web link.

For thousands of years, people in country areas burned the dried dung or droppings of animals as fuel, including dung from horses, cows and chickens.

How do BIOMASS AND BIOGAS burners work?

Biomass can be burned in solid form for hot water, like coal or similar fuel, by adding it in small amounts from a storage container called a hopper. Biogas is burned in the same way as natural gas (see page 10)—the two are similar in many ways, both containing mostly methane. The size of the gas jet or nozzle is important, and so is the air flow for the flame. The methane must mix with the correct amount of air to burn steadily, otherwise it might go out—or explode.

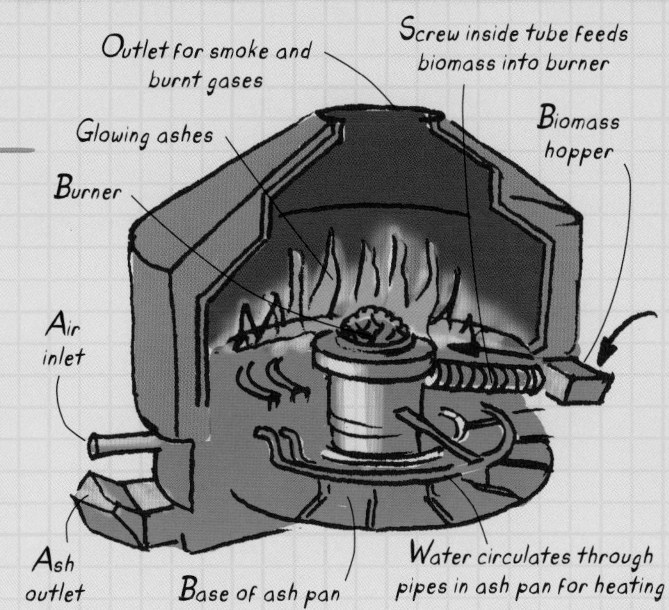

Cap The cap is sealed to keep out air, but it can be removed when the tank is emptied, cleaned or repaired.

Outlet Methane and other gases are led away along a pipe to a storage tank or container. Here they can be burned directly. Or a pump can compress or squeeze the gas into metal pressure cylinders, to be taken away and burned elsewhere later.

Dried elephant dung is especially good for burning at a slow, constant temperature—ideal for cooking.

Product The well-rotted and fermented digestate is removed at intervals and used as manure to spread on fields (depending on local regulations about recycling human and animal wastes).

Contents The biomass slurry or digestate is rotted and decayed naturally by moulds or fungi, bacteria and similar living things. These organisms recycle nutrients in nature.

SOLAR PANELS

Many forms of energy we rely on every day come from our nearest star, the Sun. They include its light to see by, its warmth and the wind turbines and hydroelectric dams that generate electricity. Another way to produce electricity is to harness sunlight directly by the electronic devices known as photovoltaic or PV cells. These are grouped together on sheets and usually known as "solar panels."

Eureka!

The first solar cell was made in 1883 by Charles Fritts. It converted only about one-hundredth of the light energy falling on it into electrical energy.

A typical photovoltaic cell produces about 14 milliwatts of electricity. This is enough for a small electronic device such as a calculator.

The latest ATF (advanced thin film) photovoltaic cells are 25 percent efficient, which means they convert one-quarter of the light energy hitting them into electrical energy.

How do SOLAR (PV) CELLS work?

In a photovoltaic cell are two layers of silicon. In one layer, mixing or "doping" with tiny amounts of boron gives the silicon "holes" where negative electrons should be, making the layer positive, the P-type layer. In the N-type layer the silicon is doped with small quantities of phosphorus, which make it negative, with extra electrons. When light energy hits the boundary between the two layers, it knocks the negative electrons from the N to the P side. This happens to millions of electrons every second and the moving electrons set up the electric current.

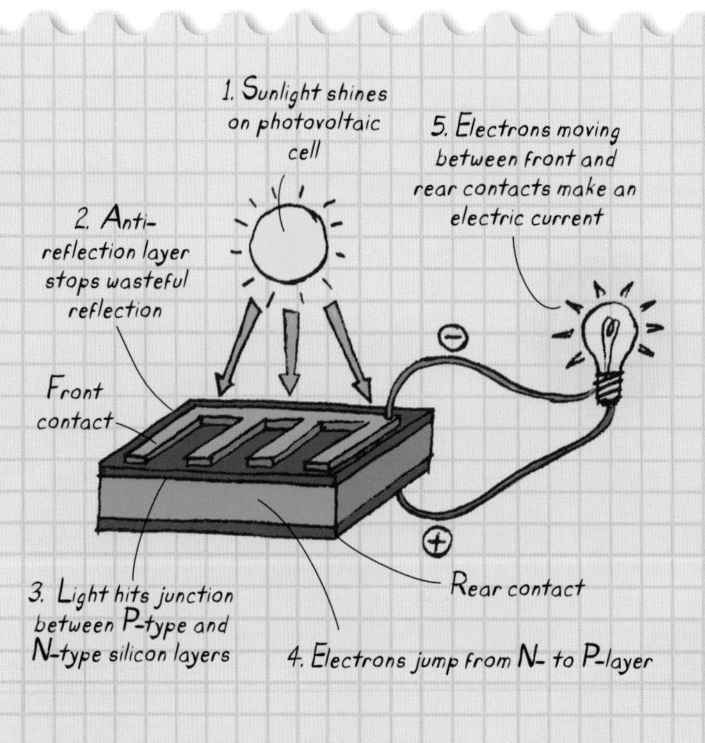
1. Sunlight shines on photovoltaic cell
2. Anti-reflection layer stops wasteful reflection
Front contact
3. Light hits junction between P-type and N-type silicon layers
4. Electrons jump from N- to P-layer
Rear contact
5. Electrons moving between front and rear contacts make an electric current

Sunlight

Electron jumps to next atom

Nucleus

Electron

What is ELECTRIC CURRENT?

Electricity is a flow of electrons. These are particles that make up the outer parts of atoms. They whizz around the central nucleus of the atom in layers or "shells" at different distances from the nucleus. In some substances, mainly metals, the outermost electrons are easily knocked from one nucleus to the next, causing a current to flow.

What next?

New materials and substances for solar cells, and mass producing them in greater numbers, will make solar power cheaper. But it will probably never become economical in cloudy places.

Carry out simple experiments and learn more about solar power by visiting www.factsforprojects.com and clicking on the web link.

If the Sahara Desert was covered in solar panels, it would provide about 40 times the amount of electricity used by the whole world.

Frame The small, individual photovoltaic cells are protected by an outer frame and joined together in a grid network as a photovoltaic array or "solar panel." Several panels connected together make bigger, more usable quantities of electricity.

A planned 32-hectare solar park in Germany, with 57,000 photovoltaic panels, should provide enough electricity for 3,000 homes.

The 1 MW "solar furnace" thermal power station at Odeillo, France

Thermal Solar Power

In addition to light, another way of using the Sun directly as an energy source is to trap its thermal or heat energy, by the clever design of houses and other buildings (see page 34). Another option is the thermal power station. Here the Sun's infrared or heat rays are reflected and concentrated by many curved mirrors into one small collecting area, producing a temperature of 5,432 degrees Fahrenheit (3,000 degrees Celsius). This is more than hot enough to boil water into steam for turbines and generators.

FUEL CELL AND ELECTRIC MOTOR

Will "clean" electric cars ever take over from petrol and diesel ones with their dangerous greenhouse gas exhaust fumes? One problem is that the batteries needed to store enough electricity to power a car for a day are bulky and heavy. The fuel cell is a portable generating device small and light enough for use in cars. It has no moving parts to wear out and makes a continual flow of electricity from the simplest substance known—hydrogen gas.

Eureka!

Scientist and lawyer William Grove developed the idea of the fuel cell in the 1830s. He called it the "gas voltaic battery." Grove also invented a new kind of chemical battery that produced a much higher voltage than others of the time. But it gave off harmful fumes and was soon abandoned.

What next?

If fuel cells catch on in cars, then we will no longer fill up with petrol or diesel, but with hydrogen gas. However changing all the refuelling stations to hydrogen would probably take at least 20 years.

The first practical use of fuel cells was in the US Gemini spacecraft in the mid 1960s.

Most fuel cells run on hydrogen as fuel, with oxygen from the air. But there are versions that use alcohol as fuel, or natural gas, or even the poisonous gas chlorine.

How do FUEL CELLS work?

Pass an electric current through water, between a positive contact or anode and a negative cathode, and the particles of water (H_2O) split into hydrogen and oxygen. This is electrolysis of water. A fuel cell does the reverse, combining hydrogen and oxygen to make water and produce electricity. The anode and cathode are separated by a PEM (proton exchange membrane). As hydrogen flows in, the electrons in its atoms (see page 31) separate at the anode and flow away along a wire as the electric current, finally returning to the fuel cell cathode.

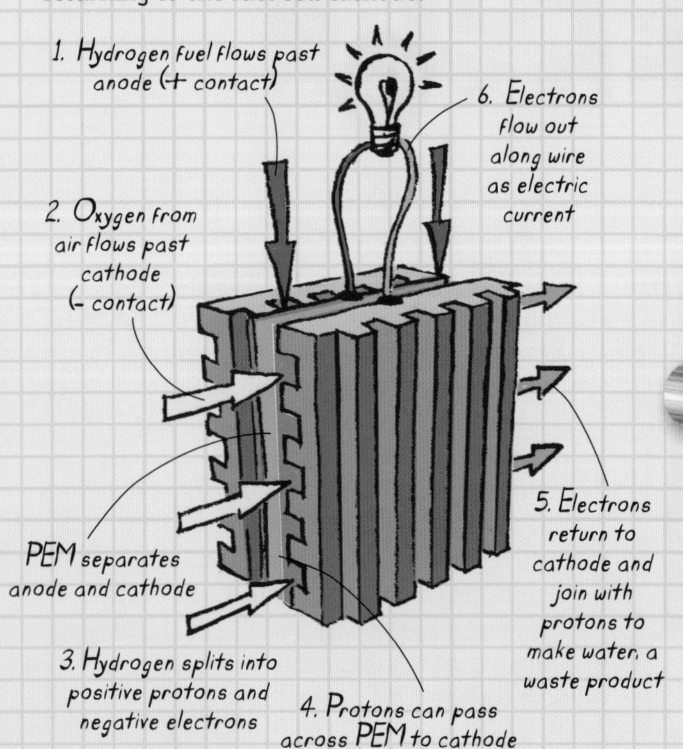

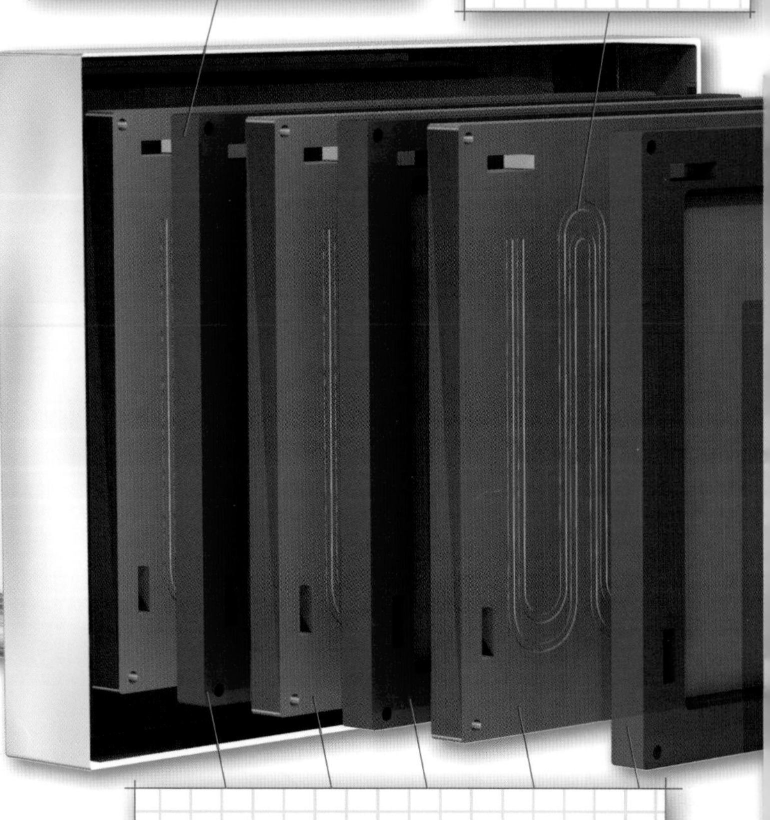

MEA The membrane electrode assembly houses one PEM (see left) unit of the "stack." Many units are butted end to end for more electric current.

Gas flow plate Hydrogen and oxygen flow easily to the exchange membrane along these grooves.

Stack A single fuel cell produces only about 0.6 volts of electric current. So many of them are linked together, as with photovoltaic cells, into a fuel cell stack that generates usable amounts of electricity.

Watch an animation of a fuel cell in action and take part in a quiz about motors by visiting www.factsforprojects.com and clicking on the web links.

Brushes Smooth carbon-based pads called brushes press on the commutator and pass electricity to it.

Magnet A powerful permanent magnet around the armature produces a non-changing magnetic field, inside which the armature rotates.

Commutator This set of contacts spins around with the armature. It receives electricity from the brushes and feeds it through the wire coils, changing the direction of electricity in the coils as they turn.

Armature Many separate sets of wire coils make up the armature. Their magnetic fields alternately push or pull against the stationary field of the permanent magnet to make the armature spin.

ELECTRIC MOTOR

FUEL CELL

An electric motor is very efficient. It turns more than 80 percent of the energy put into it, as electricity, into movement energy. This compares with . . .

. . . less than 25 percent efficiency in a typical gasoline engine.

MOTORS EVERYWHERE!

The most effective way to turn electrical energy into moving or kinetic energy is the electric motor. It works in the reverse way to an electric generator, using the electromagnetic effect (see page 15). The wire coils are electromagnets that interact with the magnetic field of the permanent magnet around them, causing the coils to spin on a shaft. Electric motors have thousands of applications, from huge electric trains to electric toothbrushes and computer hard disc drives.

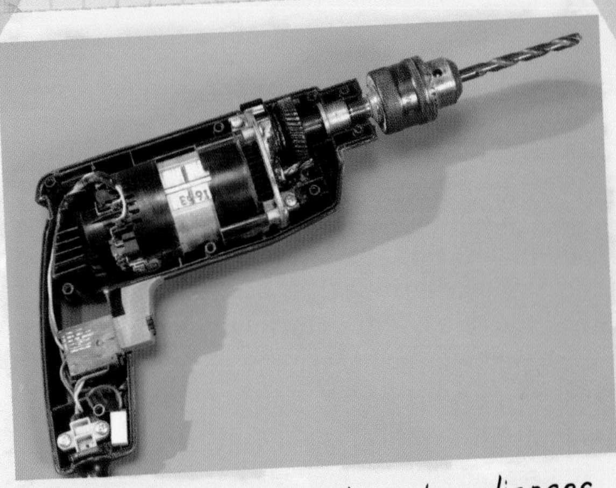

Many everyday tools and appliances rely on electric motors

ENERGY-SAVING BUILDING

Burning coal, oil and other fuels is causing great changes to our world. The greenhouse gases produced mean global warming and climate change. To reduce the damage, we must design buildings that use natural, sustainable forms of energy, and use them more efficiently. This includes heat, light, electricity, and saving water and other resources.

Eureka!

The solar stove uses a dish-shaped shiny metal reflector to concentrate the Sun's heat onto a pot, pan or oven. The first solar box oven was invented in 1767. Modern versions produce temperatures of more than 752 degrees Fahrenheit (400 °C).

A bath full of drinking water takes the same amount of energy to treat and purify as a flat-screen television left on for about 24 hours.

A typical house loses about one-quarter of its heat through the roof, one-third through the walls, one-fifth through the windows, one-tenth through the doors and the rest into the ground below.

Large triple-glazed windows facing the midday sun are one of the best ways of saving energy in a house.

Pipe loops of a ground source heating system are ready to be buried

Heat from the ground

Every day the sun shines on the ground and its heat soaks into the soil, plants and rocks. Ground source heating systems gather this heat and carry it into a building, to help warm rooms and produce hot water. The system has loops of pipes buried in the ground around, through which water circulates. It takes in warmth from the ground, which is extracted by a heat pump in the building. The heat pump works in a similar but opposite way to a fridge, moving heat one way and cold the other way.

Chimney Internal ducts from the chimney carry warm air around the dwelling to heat the other rooms. In hot weather they work as cooling ducts.

Sun-trap windows Windows that face the sun allow in its rays, which are then converted into heat, as in a greenhouse. This means even the winter sun has a warming effect.

Wind turbine Extra electricity produced by a domestic wind turbine, and not needed by the house, can be fed into the main distribution grid.

For lots of useful energy-saving tips visit www.factsforprojects.com and click on the web link.

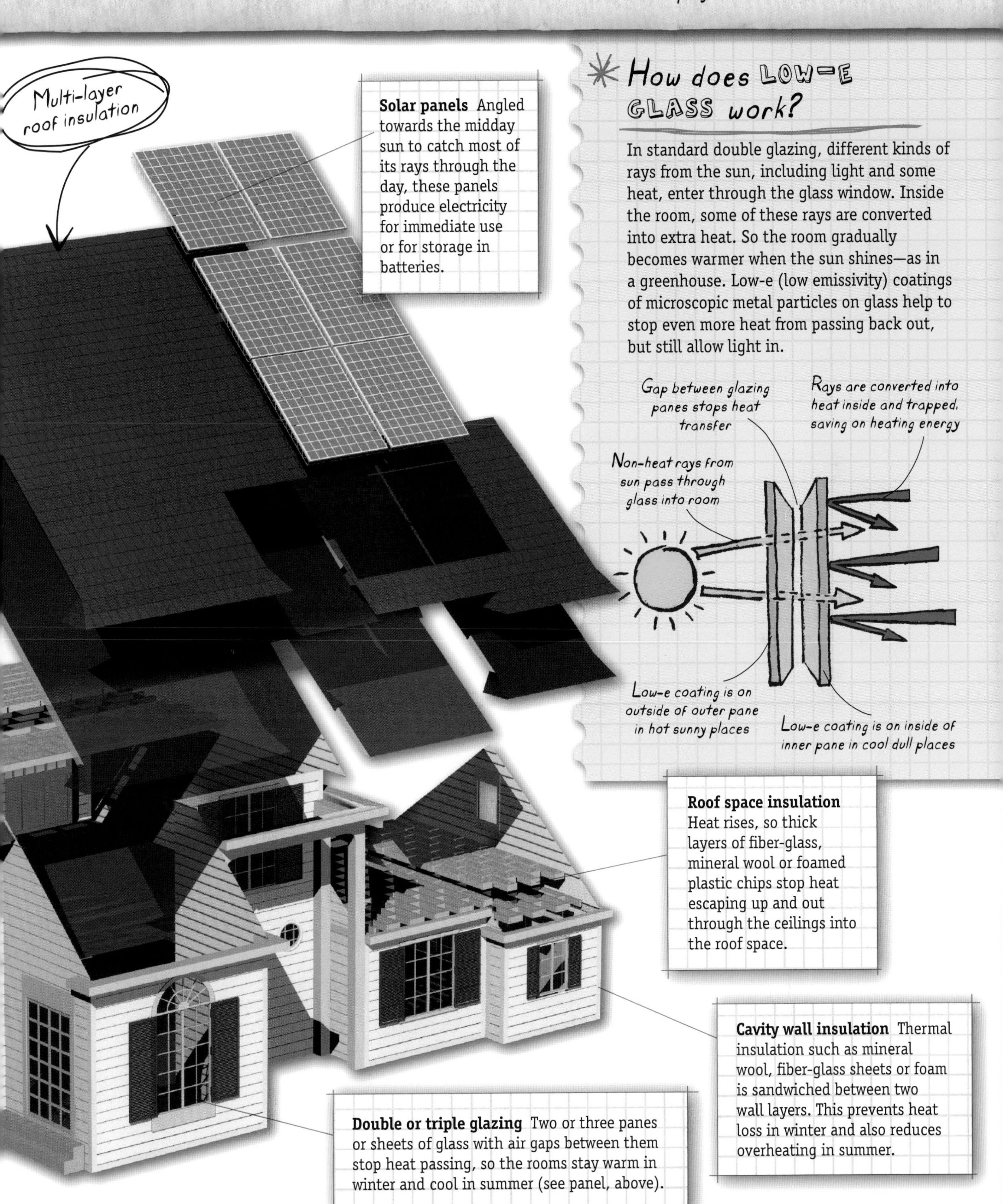
Multi-layer roof insulation

Solar panels Angled towards the midday sun to catch most of its rays through the day, these panels produce electricity for immediate use or for storage in batteries.

How does LOW-E GLASS work?

In standard double glazing, different kinds of rays from the sun, including light and some heat, enter through the glass window. Inside the room, some of these rays are converted into extra heat. So the room gradually becomes warmer when the sun shines—as in a greenhouse. Low-e (low emissivity) coatings of microscopic metal particles on glass help to stop even more heat from passing back out, but still allow light in.

Gap between glazing panes stops heat transfer
Rays are converted into heat inside and trapped, saving on heating energy
Non-heat rays from sun pass through glass into room
Low-e coating is on outside of outer pane in hot sunny places
Low-e coating is on inside of inner pane in cool dull places

Roof space insulation Heat rises, so thick layers of fiber-glass, mineral wool or foamed plastic chips stop heat escaping up and out through the ceilings into the roof space.

Cavity wall insulation Thermal insulation such as mineral wool, fiber-glass sheets or foam is sandwiched between two wall layers. This prevents heat loss in winter and also reduces overheating in summer.

Double or triple glazing Two or three panes or sheets of glass with air gaps between them stop heat passing, so the rooms stay warm in winter and cool in summer (see panel, above).

LOW-ENERGY TRANSPORT

Cars are common for going to school and the shops, fun trips and holidays. But they use up fast-disappearing fossil fuels and they give out dangerous exhaust fumes and greenhouse gases. To save both energy and the environment, we need cleaner, greener ways to travel, such as public transport like buses and trains, also electric vehicles, bicycles and our own feet.

Eureka!

The first maglev train service to carry passengers ran in Hamburg, Germany in 1979 at an international transport exhibition. The track was 2,953 ft (900 meters) long and more than 50,000 passengers took the trip.

What next?

Sadly hover-cars are still in the realm of science fiction. No known power source could make them small, safe and powerful enough, until perhaps scientists discover some kind of anti-gravity drive.

Magnetic levitation means the train uses magnetic forces to levitate—float or hover just above the track, without touching it.

Even the best-designed wheeled train has friction or rubbing due to its wheels, axles and other mechanical moving parts. Maglev trains do not suffer friction and so can reach greater speeds. A planned Japanese maglev railway line should have trains swishing along at 311 mph (500 km/h).

ELECTRIC VEHICLES

Several types of more energy-efficient, less polluting vehicles are finding their way onto the roads. Electric vehicles (EVs) have only an electric motor to drive the road wheels. They store their electrical energy in rechargeable batteries that need to be plugged into an electricity supply when discharged. A fuel-cell car makes electricity for its electric motor as it goes along (see page 32). Hybrid vehicles run on an electric motor and batteries, and also have a small petrol engine that can drive the road wheels when the batteries run down. The engine may also recharge the batteries while the car is on the move.

Electric taxis are quiet and make no exhaust fumes

Propulsion coils These wire coils (blue) carry changing electric currents. So their magnetic fields change and interact with the magnetic fields of the train's propulsion coils, pulling the train along from the front and pushing it from behind.

Levitation coils A second set of coils (red) in the guideway sides provide the magnetic forces that interact with the train's levitation coils, to make the train rise above the guideway by a few inches.

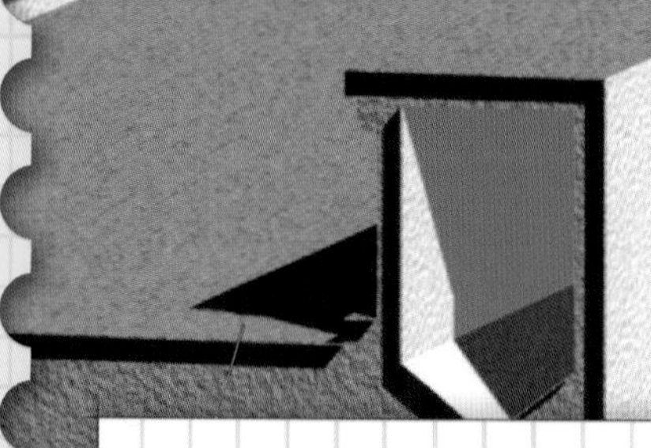

Supply rail The central rail supplies electricity to the train by means of sliding contacts on the train's underside.

Guideway surface The surface of the track or guideway on either side of the central supply rail is kept clear and smooth, in case the auxiliary wheels need to be used.

Discover everything there is to know about maglev trains by visiting www.factsforprojects.com and clicking on the web link.

MAGLEV TRAIN

Some types of maglevs use superconducting electromagnetic coils. If the coils are kept extremely cold, electricity keeps circulating in them even when the supply is switched off.

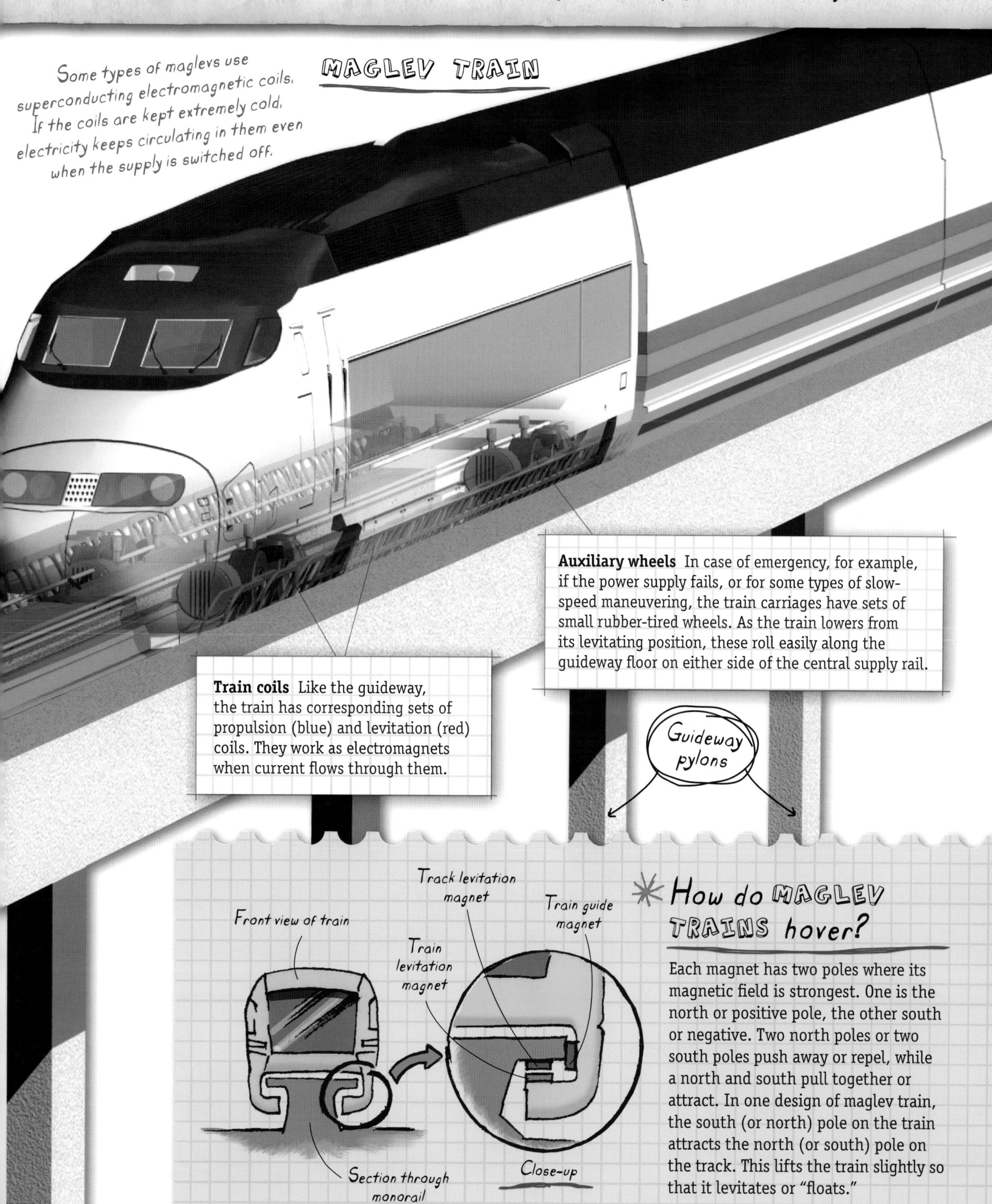

Auxiliary wheels In case of emergency, for example, if the power supply fails, or for some types of slow-speed maneuvering, the train carriages have sets of small rubber-tired wheels. As the train lowers from its levitating position, these roll easily along the guideway floor on either side of the central supply rail.

Train coils Like the guideway, the train has corresponding sets of propulsion (blue) and levitation (red) coils. They work as electromagnets when current flows through them.

How do MAGLEV TRAINS hover?

Each magnet has two poles where its magnetic field is strongest. One is the north or positive pole, the other south or negative. Two north poles or two south poles push away or repel, while a north and south pull together or attract. In one design of maglev train, the south (or north) pole on the train attracts the north (or south) pole on the track. This lifts the train slightly so that it levitates or "floats."

GLOSSARY

Anode
The positive (+) electrical contact that is part of an electrical pathway or circuit.

Atoms
The smallest pieces or particles of a material or substance, each made of a central nucleus with electrons going around it.

Bearing
A part designed for efficient movement, to reduce friction and wear, for example, between a spinning shaft or axle and its frame.

Biofuels
Energy-rich substances (fuels) that are made from recently living things, such as wood, the gases from decaying or rotting plant and animal matter, plant oils and animal droppings.

Biomass
Living or recently-living matter from plants and animals that can be made into biofuel (see above).

Blade
One of the long, slim parts of a propeller (airscrew), helicopter rotor or turbine.

Biogas burner

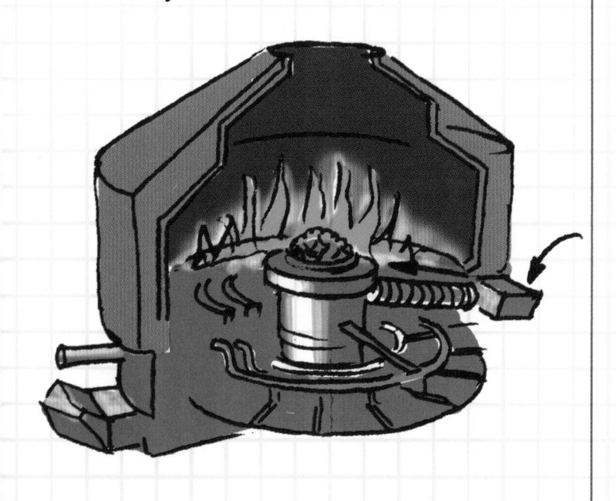

Generator

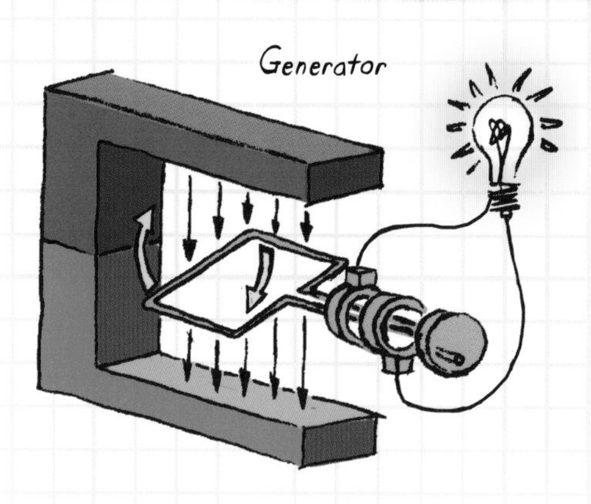

Cathode
The negative (-) electrical contact that is part of an electrical pathway or circuit.

Dope
In electronics, to add tiny amounts of a substance, the dopant, to large quantities of another one, to alter the way electricity flows.

Electrode
An electrical contact, or part of an electrical pathway or circuit, that is either positive (+), the anode, or negative (-), the cathode.

Electrolysis
To split apart or break up particles of a substance, usually a fluid (liquid or gas), by passing electricity through it.

Electrons
Tiny particles inside atoms with a negative electrical charge or force, that move around the nucleus of the atom. Lots of electrons moving along between atoms make an electric current.

Fossil fuels
Fuels that were formed over a very long time, usually millions of years, from once-living things that were trapped in rocks and changed by fossilization. The main fossil fuels are coal, oil (petroleum) and gas (natural gas).

Friction
When two objects rub or scrape together, causing wear and losing movement energy by turning it into sound, heat and other forms.

Fuel
A substance that contains lots of energy, which is usually released by burning. In some cases fuels are used in other ways—in a fuel cell, hydrogen fuel is converted into electricity.

Generator
A machine that changes kinetic or mechanical energy—the energy of movement—into electricity.

Geo-
To do with rocks and the Earth. For example, geothermal energy is heat (thermal) energy from rocks deep below the surface of the Earth.

Gravity
The natural pulling force or attraction that all objects have, no matter what their size. Bigger objects have more gravity than smaller ones.

Hybrid vehicle
A vehicle, such as a car, with more than one method of moving or propulsion, such as a gasoline engine and an electric motor.

Hydro-
To do with water. For example, hydroelectricity is generated from the energy of moving water.

Maglev train

Fuel cell

Incinerate
To burn something thoroughly, usually at very high temperature, so that almost nothing is left except ashes.

Infrared
A form of energy, as rays or waves, which is similar to light but with longer waves that have a warming effect.

Insulate
To prevent something, usually electricity or heat, moving or to greatly slow its movement. Electrical insulators such as plastic and glass stop electricity flowing, and thermal insulators such as fibre-glass and foamed plastic slow the passage of heat.

Motor
A machine that converts electricity into mechanical power to drive another device.

Nozzle
A cone- or trumpet-shaped part or device where gases come out, as in a rocket engine or gas burner.

Nucleus
The central part of an atom, made of small particles called protons and neutrons, and around which even smaller particles called electrons move.

Photovoltaic (PV) cell
A small electronic device that changes the energy of light into electrical energy.

Pitch
The angle of a turbine or propeller blade compared to its direction of movement.

Power
In everyday life "power" often means electricity, as in a power station, which generates electric current. "Power" is also sometimes used with a similar meaning to force or pressure. As an official scientific term "power" is the rate of changing or transmitting energy, or the rate that objects are moved or work is done, in units of time. It is measured in watts or horsepower.

Radioactive
Giving off or emitting certain kinds of rays and/or particles known as radiation.

Solar
To do with the sun.

Solar panel
A large flat device, which is sometimes called a solar array, that converts sunlight directly into electricity using lots of PV (photovoltaic) cells.

Thermal
To do with heat energy and temperature.

Thrust bearing
A bearing for a shaft or axle that is designed to cope with other forces as well as the usual spinning movement of the shaft or axle.

Turbine
A set of angled fan-like blades on a spinning shaft, used in many areas of mechanics and engineering, from pumps to jet engines and the turbogenerators in electricity-generating power stations.

Turbo
An engine, pump, generator or similar device that works using a turbine.

Winch
A winding mechanism that turns or reels in a rope or cable, slowly but with great force.

Yaw
When an aircraft steers or turns to the left or right, or when a machine swivels around on an upright shaft, as in a wind turbine.

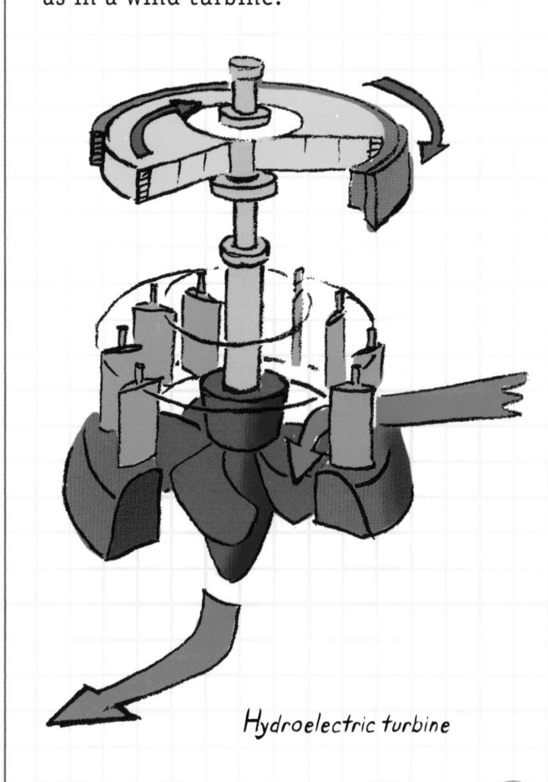
Hydroelectric turbine

INDEX